Dear Colleague.

This book is a cummulation of 30 years of Cardiac Surgical practice. I hope you may find it helpful in developing your own career. I wish you the best of luck in the completion of your training.

Best wishes

Jacques [illegible]

Lisa —

All the best

Doug Wilson

Medicine for Life:
A Practical Guide for Success

Jacques G. LeBlanc M.D., F.R.C.S.C.

Copyright © 2015 Jacques G. LeBlanc M.D., F.R.C.S.C.

All rights reserved. No part of this book may be reproduced, stored, or transmitted by any means—whether auditory, graphic, mechanical, or electronic—without written permission of both publisher and author, except in the case of brief excerpts used in critical articles and reviews. Unauthorized reproduction of any part of this work is illegal and is punishable by law.

ISBN: 978-1-4834-4186-3 (sc)
ISBN: 978-1-4834-4185-6 (e)

Because of the dynamic nature of the Internet, any web addresses or links contained in this book may have changed since publication and may no longer be valid. The views expressed in this work are solely those of the author and do not necessarily reflect the views of the publisher, and the publisher hereby disclaims any responsibility for them.

Any people depicted in stock imagery provided by Thinkstock are models, and such images are being used for illustrative purposes only.
Certain stock imagery © Thinkstock.

Lulu Publishing Services rev. date: 12/16/2015

Contents

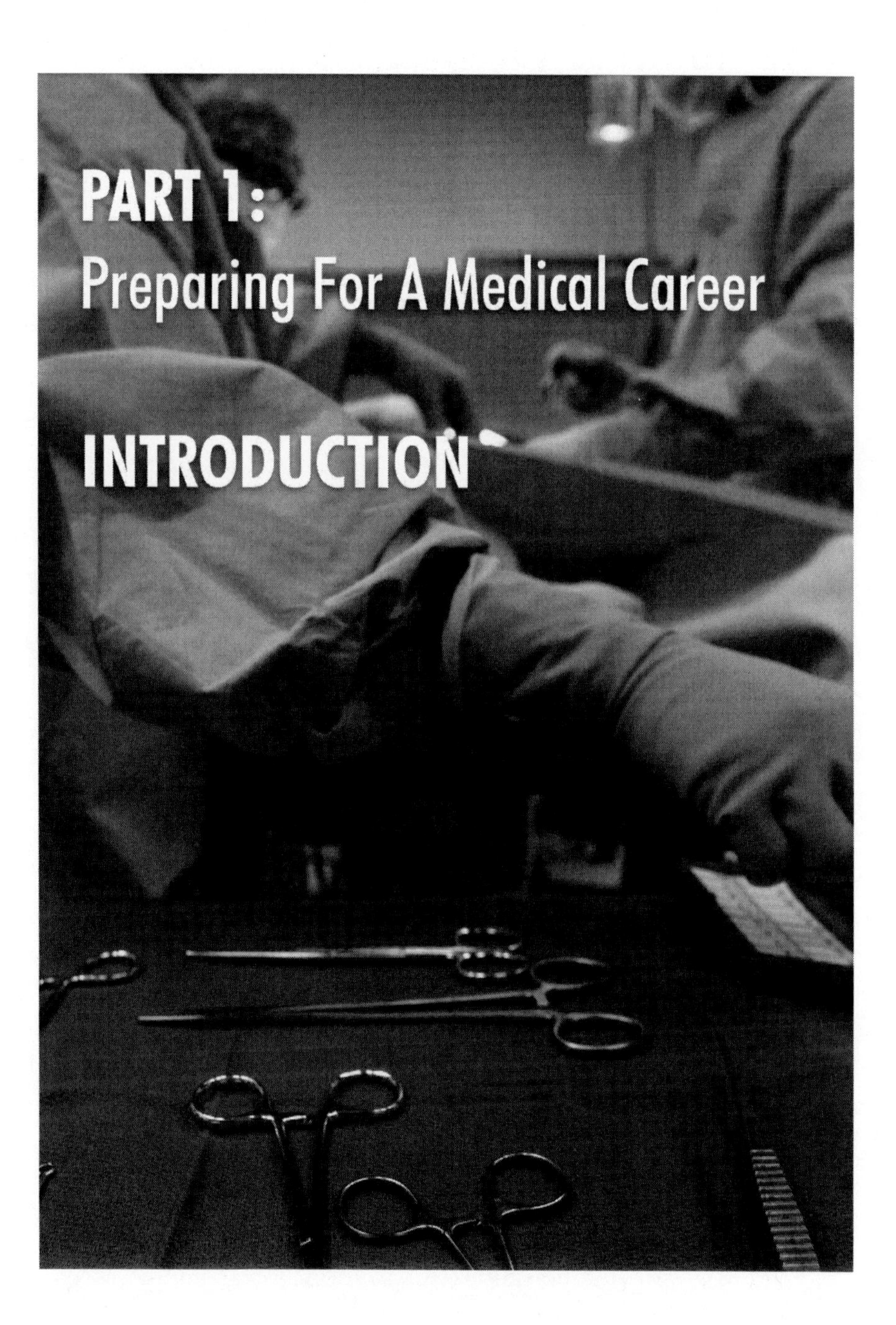

PART 1: Preparing For A Medical Career

INTRODUCTION

The inspiration for this book came from my 45 years of experience in the medical field. My journey started when I was trying to find the ideal career and then I progressed through all the various stages of searching, discovering, experiencing, planning, managing, retiring and reflecting back.

I discovered that sharing my knowledge and the lessons that I have learned along the way could be very helpful for others. Simply put, when I started my career in medicine, it was often the insight given to me by my peers that best enabled me to move forward.

Volumes have been written on each subject mentioned in this book, but the focus here is to put all of this valuable information into one book. The content will cover the full spectrum of developing and establishing a career- from starting your career in medicine all the way to retirement. This book will help you smooth out the bumps you will find along the way and move you onto the path to a successful career.

At around 16-18 years of age, you may be unsure of what your career should be. You may find that you are able to determine your career after high school, college or university. To find the best career for your life's path, you may discover that you still have a lot of educated guessing to do to find the right fit.

Along the way, you may add an extra year (a year of research for example), you may take a part-time job or spend a year working. This additional experience may prove to be very valuable for you later on. You may also learn that the degree does not necessarily make the person and the path to a career is not a linear route.

Like most residents and fellows, I was faced with several choices: medical versus surgical, general surgery versus subspecialty, to participate in a fellowship or not.

I was surprised to find that there was no single comprehensive source of information to help me, and my decisions came mainly from trial and error.

Finding a job can be quite stressful and not as simple as you might think. Just because you have a medical or subspecialty degree, does not mean that you will be given a job offer. I found that many residents often ask informal networks, medical associations, friends, colleagues, mentors, to help them through the job search process. While recruiters are readily available, they have their own difficulties and issues. This option may not be as helpful as you would hope for.

After Residency

After the residency training, there will also be more choices not only to find a job, but a wide range of practice options available: solo medical practice, joining an existing practice, a hospital-based practice, a private managed care group, a university teaching or research position.

Whatever choice you make, it will require a minimum of 6 months to get organized and to take those steps from your residency to practice. Because of the demands of your studies followed by the long exhausting process of starting a practice, this is an ideal time to take a holiday before embarking on your full time practice.

Type of Practice

The type of practice you choose will also have its own issues. The first challenge is to have a contract with your colleague and your employer that protects you, that is explicit about income, income sharing, salary raise scale, workload, benefits, expectations, conditions of termination and so on. It is a complex process and one should have a lawyer or at least a trustworthy friend/colleague to review the contract and provide feedback.

Establishing a solo practice can be very stressful. Finding the proper location, signing a lease, obtaining a bank loan, buying furniture and equipment, and hiring staff can all be part of the process.

Joining an established practice will bypass most of these steps, but you will have an established environment that you will have to adapt to, to be with a group of people that you do not know, to work under some constraints you may find difficult. Ultimately you will have to find a balance and decide if this is the best option for you.

Finances

It is well known that physicians, especially new graduates, are not always adept in financial planning, management and legal matters. It is not the objective of the residency program to educate their young trainees in these matters. But there are multiple tools and sources of support available to help you: the Canadian Medical Protective Association (CMPA) for practice insurance and, for legal matters, the Royal College of Physicians and Surgeons of Canada (RCPSC) for continuous education, for management courses, and a variety of other services.

Banking is a necessity and a private banking service banker can organize loans, lines of credit, secure funding for diverse transactions, mortgages and the like. The cost is minimal and the time, which can be saved, is very valuable.

Lack of financial instruction is the greatest deficiency of the medical establishment. This book will not tell you how to get rich because this should not be the goal. This book will provide practical, common sense advice on how to obtain your long-term financial stability and to be able to plan for a solid retirement.

Key people can provide support to your practice and your personal life. These pivotal people will establish the foundation for a successful career in medicine.

A Secretary

It is a great advantage if you have a secretary who can work with you for the long term. He or she can manage all aspects of your office practice and make your life easier, more efficient and ultimately more relaxed. Hiring an efficient secretary will be an important task and this book will provide guidance on the process.

Travel

The Internet provides the opportunity to plan holidays easily. Access to information on any country, any type of accommodation, a variety of flights, etc. is all available at the tip of the finger.

One small consideration when perusing all of this information is that it can take up your valuable time. In a busy practice, it may be time consuming to organize a holiday, book hotels and flights. Mistakes can easily be made, which may lead to difficulties during your trip. The price of a travel agent is built into the bookings and is a very small cost in comparison to the time it would take you to make the bookings yourself. This will save precious time that could be spent working at your busy practice.

Finding a Balance

Who said life is easy? Physicians and health care workers are typically trained in a hospital setting. This environment provides them with the experience with a remarkable constellation of pathologies, seriously ill patients and death. In fact, as our health care has evolved to more outpatient care and care at home, the severity of patient's illness in hospitals has intensified and increased the daily stress of our work.

Residency training programs have strict on-call guidelines; these may not apply to a solo practice or the group practice or hospital practice you decide to join. As you become more established, your clinical workloads usually increases, responsibilities change and you may become involved with your practice's management or even hospital administration.

Substantial time will be spent working. You may not have anticipated the long work hours. Along the way in your life your family responsibilities may change with the birth of your children, looking after a young baby, their schooling and the after school activities of your children. You could be potentially juggling your husband/wife and a career. You will need to work at continuing to evolve with the multiple aspects of your work, finances, and family, physical and spiritual life.

My hope is that this book reduces some of the practical stresses of setting up your business and managing life so that you can focus on the aspects of your profession that you enjoy the most.

PART 1:
Preparing For A Medical Career

CHAPTER 1:
Deciding To Be A Doctor

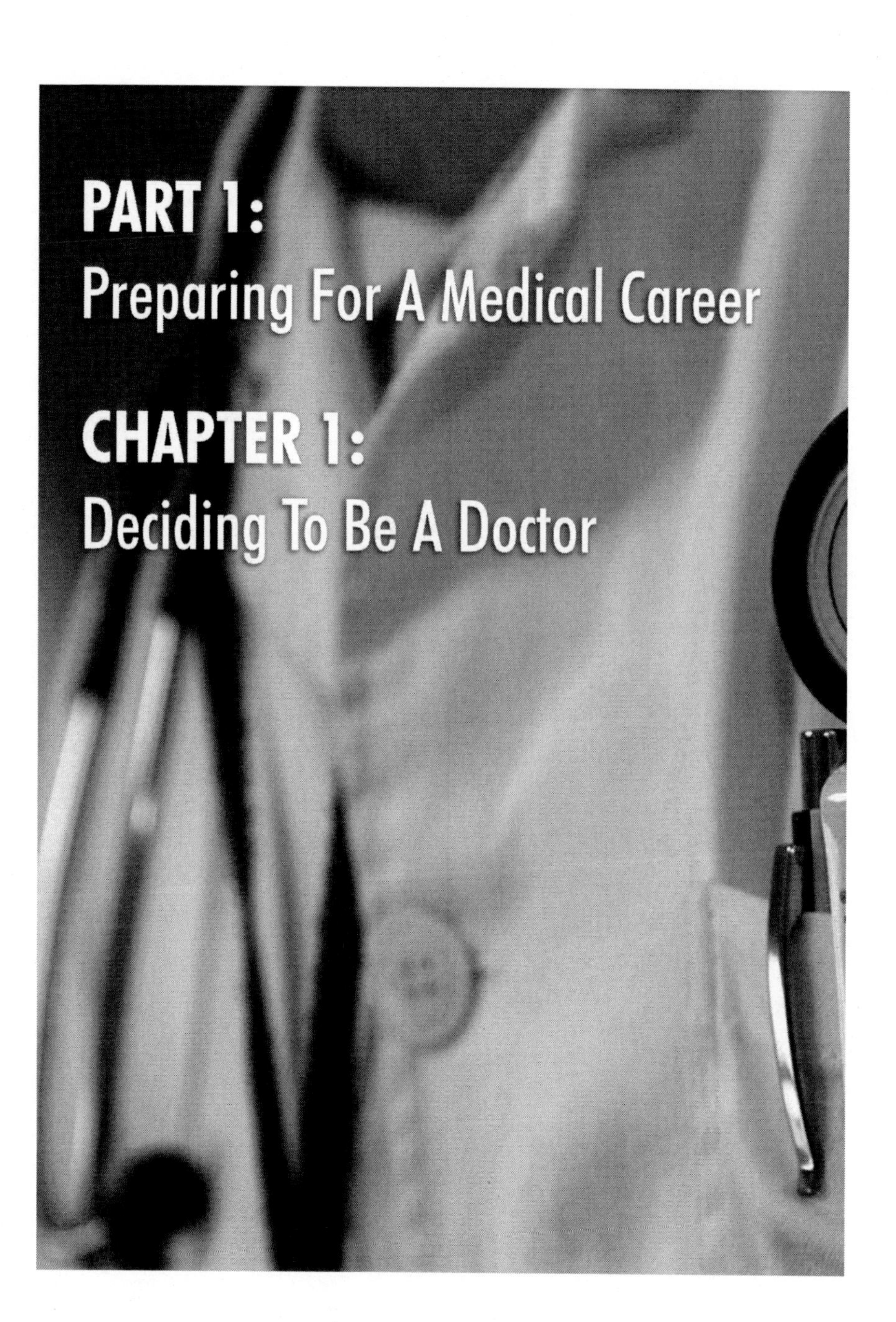

I hear from my nieces and nephews who are in high school that grade 11 and 12 are the determinant years. Physical growth and maturity are taking place at a rapid pace, interests are changing and new interests are forming. Some teenagers already know that they want to study medicine.

"Alfred knew he wanted to be a doctor since he was 15 years old. He is now, after his first year of medical school, considering a surgical career. This is after spending summers doing research on becoming a surgeon."

"Kate has wanted to be a nano-technology engineer for several years: although this is not a medical career, she is determined and has applied to an international university."

"Others may change their focus after summer work. Monique wanted to be a nurse but after a summer of medical research as a student, she decided to do medicine and has carried on to medical oncology."

"Tony raised money with other students at his school and went on an organized trip through his school to Central America to build houses in a poor area, to shadow doctors in a countryside small hospital and has had such an awakening experience that he is orienting his undergraduate courses to medicine."

Others do not know and may need to do an undergraduate degree and perhaps get working experience before making a final decision on their career. No matter which decision you have made, you may now follow the path of medicine or health care sciences as a career.

Choosing a University for an Undergraduate Degree

Even in high school, there are several steps to take in order to prepare for undergraduate school. Some classes may not be interesting, others appear irrelevant to what you aspire to do, but the most important step will be to work toward achieving the highest marks possible.

According to the, *MacLeans Canadian Universities Guide Book 2013* and to the, *MacLeans Canadian Universities Guidebook 2014,* most universities will request marks in the mid to high 80's. It is not very difficult to achieve these high marks with hard work and perhaps extra summer courses if required. If you can achieve these high marks, this will allow you to enter the university of your choice. The Canadian Universities offering medicine are ranked from 1 to 12, according to the "*MacLeans Canadian Universities Guide Book 2013*", but the book also refers to two major surveys that reveal what students think about their university experience. These surveys involved more than 37,000 students across Canada and asked dozens of questions about what they are doing in and out of the classroom. This is good information for university administrators who use this feedback from these surveys to assess the quality of their programs and services.

Results from these surveys are of good value and interest to help you decide which school is right for you - The "*National Survey of Student Engagement*" (*NSSE*) (2011) and the "*Canadian University Consortium*" (*CUSC*) (2012).

In 2012, 23 Canadian schools took part in the NSSE survey with over 22,000 students responding. The NSSE is mainly designed to measure engagement and best educational practices by measuring benchmarks such academic challenge, student interaction, learning process, educational experiences and university environment support. The CUSC survey focuses on student satisfaction from 37 universities with over 15,000 students participating.

These surveys may be very helpful in deciding which university to choose for your undergraduate program. One question I have been asked on a number of occasions is, "Does the size of the university matter?"

Thousands of young Canadians face the same question every year. The size of the university and of the community will truly define the student's experience.

There are advantages and disadvantages to both. One key difference is the number of students. Large schools like the University of Toronto and University of British Columbia have an undergraduate population of 40,000. Smaller universities mean smaller numbers of students and smaller classroom sizes, better possibility of individualized teaching and a higher likelihood of informative discussions with professors. Large universities will offer a diversity of academic options, are most likely affiliated with hospital networks, offer opportunities with nongovernmental organizations and tend to attract more academics.

There is no right answer when it comes to choosing a university. Different personalities will suit different universities.

Timing is important for your application and registration. When you have selected your choice of programs and universities, preferably around September of your 12th grade, it is a good idea to confirm the application deadlines. As soon as the applications are available and accessible on-line (usually by October) you should start your application. Many provinces have a central registration service.

The first wave of deadlines is usually in January. By then, you should have completed your applications to the different universities that you would like to apply to (remember it is a good idea to have a back up copy and to print out a hard copy of the confirmation for each application).

By March, letters of admissions will be sent out. Not all universities reply by March and you may need to wait one or two more months to receive your responses. There is usually some flexibility in the time that you have to provide your acceptance reply to a university. If you have another university that you are more interested in, the first university may wait while you wait to hear a response from the first university of your choice.

You always have the option to accept a less acceptable university program and then cancel when you receive the answer from the university of your choice. However, make sure to check the policy on tuition deposit fees. Your deposits may be non-refundable. Check with the university registration and admission/registration office regarding their refund policy.

When you have been accepted by a university and confirmed at the school, it is advisable to enquire about housing residence applications and deadlines if you would like to live on campus.

Home Away from Home

You are now at the university, most likely the first time away from home in a new environment. Most of your friends may be at other universities and/or other faculties. In a very short period of time, you will make new friends and figure out campus life. The sooner you make connections, the sooner you will feel at home and continue on with your learning.

The Sciences Curriculum may be difficult but most of the courses are mandatory for application to a medical school. A few of the courses are optional.

Some courses can be complicated and you may be disappointed with your marks and get frustrated. Just carry on, speak with your friends, professors, and family, and you will get through this initial period. Keep in mind that not all professors will communicate in a similar way. But these teachers are very knowledgeable and interesting to work with.

As mentioned previously, if you choose to attend a large university, it is more likely that there will be more students in your classes and less individual interaction with professors. Group work with student friends may replace this professor-student interaction and provide other learning opportunities.

As your undergraduate goal is to successfully apply to medical school and become a doctor, the most important step is to study hard and get the best marks possible. Although medical school requirements will place similar emphasis and importance on curricular and extracurricular activities in your application, it is still very important to maximize your chances of attending medical school by having high marks.

An undergraduate degree allows you to apply to medical schools after your third year. However, to have the best chance of being accepted to a medical school it is advisable to have a full science degree with 4 years of undergraduate schooling.

The 4th year of an undergraduate degree is important for several reasons: to obtain the highest marks possible, as the marks from your last 3 years of school tend to count the most.

Usually in your first year your marks will be a little lower than you are used to (as you are in a new place, new class, new environment, with different and more difficult subjects). Therefore, high marks in your 4th year will bump up your overall average. Completing your 4th year will allow more time to build up your extracurricular activities that are required for your successful application to medical school.

The experience you acquire from doing volunteer work, summer research projects, working part-time in a university laboratory, coaching a sports team, even working in a store or business, will be valuable in the years to come.

Another important point is to build strong relationships with your professors during your undergraduate years. You do not need to be friends with your professors and, in fact, this is ethically wrong. But it is crucial to keep in mind that you will need academic reference letters in the future.

Despite large class sizes and some shyness on the part of the student, students can still build relationships with professors through discussions, research projects, questions sessions, readings and weekly students' groups, etc. Through these activities you will find it easy to build relationships with your professors.

Finding Work

Finding work is important for two reasons: for acquiring work experience and earning a living.

Despite the busy schedule and the required number of credits to take per semester, there is time for a part-time job. A few hours of work per week during the school year can translate into a full-time summer job. Part-time can be quite varied, from working at a food chain to working in your professor's laboratory, to tutoring younger students, to coaching a sports team, to monitoring the university gymnasium, etc.

There are multiple opportunities; some can be easy to find, some may be more difficult. All it takes is for you to start asking people about potential work and, by asking about jobs, you will be on your path to part-time employment. Ask questions, knock on doors: this will help you get ahead.

Even if you are privileged and fully funded from your parents, work is not just about money. Work about learning, identifying opportunities to develop new skills, understanding how society functions, working with people/customers, and developing people skills. Studying is the reason why you are at university, but life skills cannot be acquired by reading books alone.

Life skills are acquired through life experiences. Working with people, learning how to make decisions however small, helping others, pushing the boundaries of your own mental, psychological and physical abilities will all contribute to your success. As a doctor, you will greatly value these experiences later on.

Paying for Education

Getting an education is expensive. Now is the time to learn about money. Parents believe schools teach financial principles and teachers think parents are teaching their children about finances. The truth is probably neither prepares young people on how to manage their finances.

So now there is a new added stress: learning financial skills. Factoring all the costs, from tuition programs to late night snacks, the average price of a 4-year undergraduate degree for students who do not live at home is around $80,000 ($20,000 per year)(*Macleans 2013 Canadian Universities Guidebook*).

Don't quit because of the cost: it is worth the experience. You will find the money and medical school is just around the corner.

How do you pay for your education? Many parents will do what they can to help, over 60% of students report getting less than one quarter of their annual funds from their parents (*Macleans 2013 Canadian Universities Guidebook).*

The majority of funds will have to come from somewhere else. Summer jobs are helpful but contribute an average of less than $7,000 to the budget. Most students cannot count on student loans. The student loan funding has fallen nearly to zero, from somewhere around $60,000. This amount varies somewhat from province to province.

If you have savings, own a car, or live with your parents, the success of getting a student loan is even less. If you do apply for a student loan, check with the student loan office for their deadlines.

Student loans max out at around $13,000 per year. This is short of the $20,000 you will need every year for your schooling.

There are some advantages to student loans. The government may forgive up to 40% of loan as an "opportunity grant." Another advantage is that the interest on the loan does not start until after graduation.

The main catch is that when the interest does kick in, the interest will be higher on a student loan than on a line of credit. On a student loan, the interest starts at prime plus 2% to 2.5%.

If you do not qualify for a student loan, a line of credit is a good option. The bank's rate is often as low as prime plus 1%. Your parents may be able to help you by co-signing a line of credit (as your credit history may not meet the bank's requirements). The major drawback with a line of credit is that the interest rate is charged from the day you take money from the line of credit.

Although the interest rate is currently relatively low (4 to 5%), if you get a line of credit for $20,000, using $5,000 per year, you will probably pay $2,000 in interest over 4 years (interest rate at 5%) and up to $5,000 after your graduation if you take 10 years to pay the loan.

The line of credit will cost you $20,000 to $27,000 to pay it back (depending on the length of time you take to pay it back). Try to keep the line of credit to a minimum.

Many students may think they are too busy to work. In reality, despite the number of credits taken during a semester, time is spent studying but also socializing, watching TV/video games, sleeping. With a slightly tighter time management, there is possibility to earn money without missing out on study time. The key is to find jobs with hours flexibility that will accommodate schooling.

Filling a tax return may be lucrative. For those of you who did not have a job the year before, you may still qualify as low income. If your income is below the $10,000 level, (it varies with provinces) you can obtain a GST credit (Goods and Services Tax Credit) and students can get a quarterly payment of $65 to $95.

Those who earn enough to pay taxes (more than $10,000) can deduct tuition fees, plus a certain amount per month another for the education amount and for textbooks and materials. You can find how much by consulting any accountant (your parent's accountant or a friend).

These tuition and education fees act as credits, even if they are not used on your current income tax report. The credits can be carried over to future years when you will have a higher income from summer or part-time work.

Students can transfer as much as $5,000 in tax credits to their parents, spouse or save them to use in the year after the graduation. It is like money in the bank when you start working.

To fill the gap between financial support from parents, government, student loans and part-time work, there is the scholarship option.

Scholarships and Awards

Scholarships and awards are available throughout the year and news about them pops up all the time.

Examples of scholarships are: sport, music competition, cheerleaders, school awards for academics, etc.

Applying for a scholarship is an excellent learning experience for anyone. Preparing your curriculum vitae, writing a letter describing your personal interests and achievements, learning to write in an essay format/structure, knocking on doors for reference letters, being organized and concise are all part of the scholarship experience.

Finding a scholarship is not as hard as you think. Search for awards that fit your personal profile, interests and strengths, maximizing your chances of success.

Lastly, another option is to enroll in the Canadian Armed Forces and obtain support for an undergraduate diploma and continue with medical school. You will need to enquire carefully about the terms of references and the rules. You will want to have a very thorough understanding of what the Canadian Armed Forces will require in exchange for your education. They may require a number of years working in the forces to pay for your education. It is a good opportunity if you do not have other options for funding.

A career in the Canadian Armed Forces as a doctor can provide the benefits of a pension plan when you leave the Armed Forces.

Spending can lead to large credit card bills with compounding interest. It is best to have only one credit card and pay it off each month to avoid increasing interest fees and to build a good credit rating.

Using a simple computer software budget program can help to track money coming in and money going out. Wine, beer, and cigarettes are costly. Coffee (café latte, cappuccino, and the like) can add up. Several of these per day can add up to $6 or more. If you have these for a week the total can be $42 each week or more. It is easy to lose track of the cost.

Cellular phones are the next expense that can vary greatly according to your use, your plan. Usually the minimum cost is $50 per month. Try to text more to keep the phone expense to a minimum.

Getting Around

Transportation is an issue for everyone. Public transit monthly passes are available in most cities and are the least expensive method of transportation but may be less convenient than a car. A car is costly: gas, insurance, repairs, parking, etc. Students who drive spend 4 times more than students who use public transit.

Universities once had the reputation for offering unhealthy food options. It is changing with students asking for gluten-free alternatives, vegan and other types of diets. Providing all that diversity is a huge challenge for university food providers who need to keep costs down, respect union contracts, maintain food outlets on the campus combined with providing affordable healthy food options. Universities offer meal plans with residence accommodation but you may save money by doing your own cooking. If you live in an apartment-style residence, this may be an option for you.

A tip about meal planning is to buy the smaller or less expensive meal plan so you do not have left over meal points at the end of the year. Make sure to read the agreement for the meal plan before you purchase it, so you are not surprised by any hidden fees.

Life on the Campus

Finding the right place to live is important and can be central to setting the tone for the university experience. There is a lot of learning that is not in the classroom and happens faster in these intensive communities.

Even if you have the choice to live at home, close to your home town university, living away from your parents at the university residence may be rewarding and help you ease into adult life.

It is the halfway point into the world of full-fledged adult life and responsibility. For most students, it represents a significant lifestyle change and comes with real challenges.

While many Canadian universities offer single rooms, they can fill quickly. Most first-year students will find themselves sharing a bedroom with a roommate (don't worry: there is a pairing system to ensure compatibility).

Students share space with more than a single roommate. You can be sharing a bathroom with several people, living in close proximity with 30 people, sharing kitchen accommodations and areas of relaxation (TV room, reading room, sitting room and meeting room).

The noisy people will quiet down and most students will want to keep their academic-social balance in check. The best way to ensure a successful residence experience is by planning ahead, going on-line to find the different options and even visiting the facilities before you decide to live there.

Today, there are many living options and custom-tailored residence space. New residences are being built, old ones are renovated into more spacious, and costly apartment-style residences. Suites are a spacious but an expensive option at some universities, featuring several single rooms, central living space, and a full kitchen.

Suite-style residences may leave some students slightly isolated from the group but for other students this may strike the perfect balance.

In the end, it comes down to personal preference and what you can afford. There will be a sense of community that will help you to transition to your new life and your home away from home.

Coming from high school and your home environment, you have probably had a stable group of friends for several years. Some of your friends may have gone to different colleges/universities or life endeavors. You will make new friends and you will come in contact, perhaps for the first time, with international students. It will be a broadening experience for you. Some of these students may have language difficulties and could appear unprepared, but it is all part of attending a large college/university.

You may even feel at times that students with language difficulties may take too much time or attention in your classes when a subject seems easier to you. Be patient and helpful: it will go a long way toward building new friendships and developing your own maturity.

Remaining Connected

The cell phone connectivity is integral part of life. A 2010 Pew Research Center Study *(Mobile Health 2010." Pew Research Center Internet Science Tech RSS.)* found that 90% of American students aged 18 to 29 sleeps with a cell phone next to their bed. The compulsion to stay plugged in is very strong as there is a constant stream of texts, emails, Facebook messages, Instagram pictures and whatever else to check.

People have no idea how poor their daytime functioning is because of all this disturbance and continuous connectedness with devices. Studies show sleep disruption has negative effect on memory and attention and contributes to daytime fatigue. Even if students say texts do not bother them late at night or during the middle of the night, it does not mean that their sleep patterns are not affected.

Last fall, between cell phones calls, emails and texts, college and university students were in touch with their parents on average 22 times per week, up from 13 times per week in 2007 (*The Connected Parent*: *staying close to your kids in college while letting them grow up, August 2010*).

Students who contacted their parents the most were also the least autonomous. Mothers, fathers and their children have more in common these days. The generation gap has shrunk. Children and their parents are now living a similar daily life. Parents have a peer-like relationship with their "adult teenagers". *"How to Succeed in College (While Really Trying): A Professor's Guide to Mastering What's Expected (2010).* Despite this better relationship, teenagers and young adults need to become more independent. It is part of building an adult life.

REFERENCES

1. Macleans 2013 Canadian Universities Guidebook 2013: 3-258. Print.

2. Macleans 2014 Canadian Universities Guidebook 2014: 4-130. Print.
3. www.macleans.ca/oncampus
    a. (click on university rankings)
4. National Survey of Student Engagement at University of North Dakota (NSSE), Sue Erickson, Carmen Williams,
    a. November 16, 2011
5. National Survey of Student Engagement." (2012): n. pag. Web.
6. Canadian University Survey Consortium (CUSC)." *Dalhousie University*. N. p., 2012. Web.
7. The connected parent: staying close to your kids in college while letting them grow up
   Barbara K. Hofer, Abigail S. Moore
   Free Press, a division of Simon and Schuster Inc.,
   New York, August 2010
8. "Mobile Health 2010." *Pew Research Center Internet Science Tech RSS*. N. p., 18 Oct. 2010. Web
9. "How to Succeed in College (While Really Trying): A Professor's Guide to Mastering What's expected." Jon B. Gould (2010): n. pag. Web.

APPLICATION/GENERAL INFORMATION WEBSITES:

- macleans.ca/oncampus
- electronicinfo.ca
- ontariocolleges.ca
- aucc.ca
- educationplanner.bc.ca
- ouac.on.ca
- applyalberta.ca
- bccat.bc.ca
- fedecegeps.qc.ca

SCHOLARSHIPS/FINANCIAL AIDS:

- canlearn.ca
- macleans.ca/oncampus
- studentawards.com

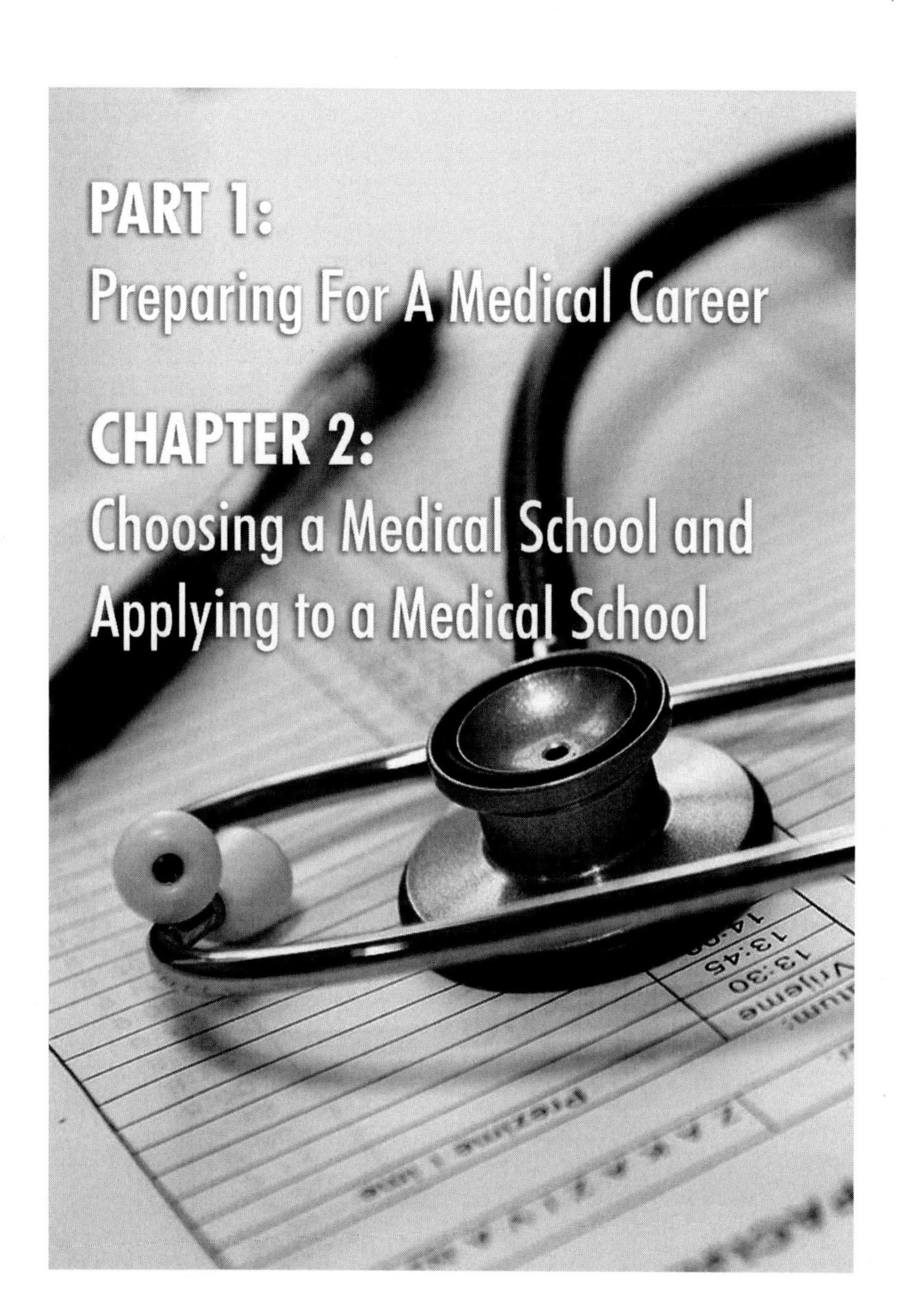

PART 1:
Preparing For A Medical Career

CHAPTER 2:
Choosing a Medical School and Applying to a Medical School

Being a doctor is a wonderful profession and offers a large variety of opportunities. You may be attracted to the profession because it is challenging, or because of the decision-making process, or because you enjoy dealing with life-threatening situations, and you find a great amount of satisfaction in saving a life.

You may enjoy treating and caring for people in need, the technicality of surgical specialties or enjoy the research opportunities. You could also be drawn to the profession because of the lifestyle and yet at the same time, you may be afraid to be a doctor or think it is not a profession for you for all of the same reasons as mentioned above.

The number of applicants to the universities, the applications requirements, and the application process may also intimidate you. Do not let these factors influence your overall decision-making process. It is competitive to get into a medical school but it is not all about the high grades and the extracurricular achievements, although those factors are important.

Being a physician is a great achievement and provides a fulfilling and exciting career. The opportunities are unlimited: you can use your hands, your heart and your mind to solve medical problems and to change lives. At the same time, you will meet thousands of people/ patients and be somehow a part of their life, part of changing and influencing their life.

Not everyone has the opportunity to work with people from all walks of life: within your community, nationally and internationally. Combined with your clinical skills, you may be a teacher, a mentor to a young colleague or trainees, an administrator for your department or hospital, or a researcher. And lastly, another benefit is the stable income you will be compensated for your work. You will also be a respected member of your community, which is a reward unto itself.

Whatever you may have heard from other students, friends, perhaps other doctors or even your own family, it is not as difficult as you might think to become a doctor. The training is long, 4 years of Medical School, 2 years of Family Practice residency training and/or 5 to 6 years of subspecialty training. There will be long hours of studying and many years at school, but other career paths have similar requirements. If you decided to become a lawyer, accountant or engineer, you would need to take the same time to attend school and training.

"Patrick did a PHD diploma in Theology. It took him 12 years of studying to obtain his PHD and he was just hired to a full professor position at the University of Toronto."

So you are not the only one pursuing your dream and consequently spending a long time on your education. Many other professionals and non-professionals do too. I do not know any colleagues, young or old, who are unhappy about their choice to become a doctor.

Choosing a Medical School

There are 17 medical schools in Canada, 130 in the United States and many more overseas. In order for you to differentiate each school, you should definitely have some ideas of what is a good medical school for you and where would you like to apply. Every person has their own set of ideas of what will be a good medical school for them, but if you know ahead of time what schools are good for you, it will make applying to and choosing one much easier.

The admissions criteria to the majority of medical schools in Canada are very similar. The admissions process for most universities is 4 years of undergraduate studies on a Science degree. The admission process is divided in Curricular and Extra-curricular achievements. The Maclean's University 2014 (*"Canada's Best Schools: 2014 Maclean's University Rankings - Macleans.ca*) report does provide information about admission curricular score per university and it hovers around 86% to 89% across Canada. Most Faculty of Medicine admissions want you to make sure you have a high GPA and can handle a rigorous medical teaching curriculum. High grades are not the whole story but a very important aspect of it. Start to work at it early.

Extra-curricular achievements will compliment the high marks that you have achieved with your studies. The list can be varied and here are a few examples:

- Sport achievements, coaching, teaching
- Research projects, teaching and student mentorship
- Volunteer work: community, national and/or International
- Public speaking, entertainment
- Music achievement
- Discovery

These extra-curricular achievements are not the most important part of your application, but it shows your journey toward medical school, your interests, passions and some aspects of who you are. You may find that you want to get involved in many activities and that is fine, or you may find you want to be more selective in your activities and concentrate your energy on a few things of interest, and that is also fine. There is no ideal personality, type of personality and no ideal extracurricular activities. Do what you feel is right for you and believe in yourself. It is a great personality trait toward being a doctor.

Many statistics are posted from each university *("2014 Maclean's University Rankings By Macleans | Theguidebook.ca."*). You may be discouraged by these stats, the level of competition and achievements like taking extra courses, being the president of your class or sports club, joining a humanitarian program or travelling the world. In reality, none of this needs to be done to a very high level. Admissions are broad-based on curricular and extra curricular activities as mentioned but also on strategies and diversity of the students. There is no set mold, or set personality, set quality or even set overall level of achievement.

There will be a set level of marks expected from each university. But this allows for a fair application process for all applicants. Good applicants can come from all walks of life. Every applicant has different strengths and weaknesses. If you are well-rounded, hard working, motivated, persistent when faced with challenges, and willing to participate in rigorous medical training and its application process, then you should not be discouraged. What counts the most is your desire to follow a medical career and to make a difference in other people's lives.

The Most Important Factors to Consider when Deciding on a Medical School

- Location
- Academics
- Finances
- Student Life
- Personal

Location

Nothing is more important to your well-being in medical school than location. In my opinion, where you attend medical school is the most important factor to consider. It will influence your academic ability, finances, education, and your personal life. Medical school is a big change for most people and to begin your medical education in a new or familiar environment will make a difference in your learning.

I don't think it matters too much which medical school you attend. Every accredited medical school will teach you quality information and the full curriculum, and prepare you for a career in medicine. Every student in Canada will take the same qualifications and board exams. In the end, no matter which school you choose to attend, you will still graduate with the same medical degree as any other doctor in Canada. If keeping in touch with family and friends is a high priority, you may want to attend a school close to home (in the same city or several hours driving distance). If you have lived at home your whole life and you are ready to see the world, this may be a good chance for a change of scenery.

Choosing a medical school is not only a personal choice, it is a decision that affects the people around you. It would be wise to examine your current relationships and friendships and see how moving away or staying will affect them. After all, a happy personal life will be important for your well-being and happiness in medical school. Likewise, if you are married or in a serious relationship, living apart for 4 years may be a significant hardship. Furthermore, location is the difference between walking to class through 3-feet of snow, cold weather or a milder climate. The climate should not make or break your decision, but having to survive through 4 years of conditions you can't stand, will surely make your time studying less enjoyable. Every small piece of life style counts.

Many people cannot stand living in a rural or small town setting. On the other hand, there are people who hate the big city life. Can you afford the cost of living in a metropolitan area? Will studying in a remote place be too boring for you? These are questions people should ask themselves. Not only will choosing the right city affect your personal life, it will help how you enjoy your education and while providing very different patient demographics.

If you decide to live at home, you can definitely save a lot of money. If you do move away, it will be a more costly option. You may need to consider living with friends or sharing your accommodation. You may want to stay on campus, at least for the first year. Not all campuses are created equally. Some are in the heart of a bustling city and others are quiet and conductive to studying. When you visit and interview at a school, be sure to take a tour of the campus and make some notes about the surroundings.

If you are planning to attend an International (Overseas) Medical School make sure you are aware of all the obstacles you will face. There will be many unique hurdles you will have to overcome, including limited clinical electives, competitive residency spots and many documentation issues. If you want to eventually practice Medicine in North America, you may need to take extra exams to meet the qualifications standards. Make sure you are aware of these issues in advance because each year many foreign medical graduates find themselves in difficult positions because they did not understand the challenges they would encounter during their studies and at graduation.

Academics

Every medical school can train you to become a competent doctor. The material you will learn is almost unanimously the same at every school. However, it is your benefit to choose a school where you can learn best. There are currently two schools of thoughts: (1) The didactic lecture-heavy traditional curriculum and (2) the "integrated" case/problem based learning (PBL).

There are strong supporters for both methods and my advice would be to figure out which works more effectively for you. If you are worried that your pre-clinical education will be compromised because of a school's curriculum, know that in every class there are good and bad students regardless of the program. In the end, all learning is self-learning; so be responsible for your own education.

There are only two Canadian schools that have 3-year programs. Both McMaster University (*Macleans 2013 Canadian Universities Guidebook, page 136-137*) and Calgary (*Macleans 2013 Canadian Universities Guidebook, page 105-106*) University program have classes throughout their summers, the total amount of class time is equivalent to a traditional 4-year program. However, there are pros and cons to completing a 3-year program.

With a 3-year program, your medical education is done faster and you can begin your residency training sooner. A 3-year program can be beneficial for students who know what specialty they want. However, there are also drawbacks to a 3-year program, including burn out and less time for electives before the residency match.

Since the program is accelerated, there are fewer chances for research opportunities normally done in the summer and not many opportunities to travel or take time off to pursue personal interests. Furthermore, many medical students do not know what field they would like to go into and many find that the extra year of school really gives them the time to think and help make that decision.

As mentioned before, the location of your school will affect the clinical exposure you receive in 3rd and 4th year. A bigger school with a larger patient catch area will provide you with potentially more rare cases and the possibility to be involved in more projects. But still, you are not at your residency training level, yet, so more or less complicated cases will not make a huge difference in your residency training. It's important to speak to graduates at other schools to see if they feel adequately trained to choose a residency program and carry on with training.

For some strange reason, there are still some schools that use a letter grading system to evaluate their students. Obviously it helps differentiate the excellent students from the poor ones, but it also comes with a lot of unnecessary stress and competition. For me, Pass/Fail has been a huge blessing that decreases hostility and builds class cooperation. Though not a must-have feature, having P/F grading is a philosophy that I believe all schools should implement.

Most schools will classify themselves as an academic research center or a primary care school. You may have an inclination for research, and if this is the case, find a school that will help you pursue these interests and make it easy for you to do so (mentors, lab space, funding) would be another thing to think about. Furthermore, if a school is strong in certain areas of medicine you are interested in, for example surgery, you will get a richer learning experience and possible increased networking opportunities in your field. Think a bit further ahead to your eventual career and try to cater your current education choices to fit your future dreams.

Finances

Tuition fees across Canadian Universities may vary slightly from university to university but it is definitely less expensive than the States. Depending on whether you go to a private or public school in the States, the tuition difference can be astronomical. Is attending an

Ivy League medical school worth the several extra tens of thousands of dollars? You should evaluate your own financial situation and decide what's best for you. Perhaps accepting that state school scholarship with a large financial package will be more valuable than any top-10 education. Graduating debt-free is something that shouldn't be passed without much thought. As mentioned before, the city or town your medical school is in, what type of place you decide to live and, what amenities and hobbies you have will factor into living costs. Every person is entitled to spend their money on whatever way they want. Just make sure that your spending makes sense for the medical school you choose and it fits your budget. The last thing you want is to be stuck paying off debts for the next twenty years or more.

Reputation

There are rare cases where accepting a position at a top school may be to your advantage. If you plan on a career in academia, it may be worth it to pay the higher tuition for the extra research opportunities offered at these schools. The value of a "name" can be easily debated by both parties. Decide where you stand on the issue and determine what would be a reasonable price to pay. Whatever you do, don't pay attention to News Rankings. They are almost always inaccurate and offer little value determining which school is right for you.

Student Life

Are you a person who enjoys large groups and getting to know everyone in your class, or would you prefer some privacy and a core group of friends for your four years of medical school? One of the things I enjoyed about my medical school years, was the class unity and closeness during my undergrad years. I don't think I would have had as much fun during this time if our class had 100 or more students. But everyone has his or her own preference.

On the interview day or your second visit to the school, were the students friendly and enjoying themselves? What were their experiences on the medical school like? Remember to ask questions about things not advertised on the school's website and brochure. Is there a life outside of the classroom? Are the students well supported, is there counseling services, maternity leave, and mentors for those who may need them? Is the social scene conductive to studying and getting along? Trust your feelings and ask yourself if you can see yourself fitting in with the school or not.

Personal

Medical school is already stressful on its own. If important pre-existing relationships become strained because of your choice of school location, it may be time to reconsider your choices. Make sure your friends and family back home support your decision. Talk to them and ask if they share your perspective.

Remember that the decision you make will take a toll on your physical and emotional health, as well as the relationships around you. If you are still unable to decide which medical school to attend, listen to your heart. Most of the time, it knows what you want, even before you know it. Did you feel a certain vibe about the school during your interview about the people, the place, the curriculum? It's important to record your impressions of each school you interview at, as it will help you make your choices quickly. In the end, you want to pick the medical school where you can be your best. There is no point in trying to fit in at a school that you don't feel comfortable with. Trust your inner voice. I have found many times that when it comes to complex and difficult decisions, your heart has an ability to simplify and accurately assess many factors.

International Medical Schools

Many Canadian students choose for a variety of reasons, to study medicine abroad. The "*Canadian Resident Matching Services*" (CaRMS) have released a comprehensive report ("*Report on Canadians Studying*

Medicine Abroad - CaRMS.") and estimate that approximately 3500 Canadian students are pursuing schooling abroad. There are 130 medical schools in almost 30 countries offering medical training for Canadians. While up to 90% of Canadians studying abroad would like to return to Canada for residency training and future practice, the process may be difficult and frustrating. Understanding the complexity may help to minimize the challenges and be successful in finding a residency training position.

The Royal College of Canada cannot and does not have the authority to review these 130 medical schools programs curriculum, their terms of reference and their accreditation process. The medical schools abroad are registered under many different agencies, governmental or universities, according to each country. Their medical education programs are diverse and their clinical components do not usually offer the autonomy and direct patient care approach promoted by the CanMEDS guidelines of the Royal College ("CanMEDS: Better Standards, Better Physicians, Better Care." *Royal College).*

CaRMS is a non-for-profit organization established in 1969 working in cooperation with the Canadian medical schools to provide a comprehensive and transparent electronic application service and a computer match for entry into postgraduate medical training throughout all the Canadian residency programs.

When applying for residency training programs in Canada from an international medical school, you have to meet the same eligibility criteria than the Canadian trainees and go through the same highly competitive process of matching. You may not meet the eligibility criteria or other aspects of the matching program and you are not ensured of success.

Please review all information provided on the website of CaRMS so you can understand the process and potential difficulties when applying for a Canadian residency training position after your medical schooling abroad. It may help your choose or pursue other venues.

Closing Remarks

Each person will have a different set of values, wants and needs. It's important to realize what may be a good decision for someone else may not necessarily be the best choice for you. It is best to first set your own criteria and boundaries you would like before going into the specific details. I would suggest starting with location and work your way down the list. I assign values to them. Once you have your priorities, start weighing the pros and cons of each school. Make a list and see what's good and bad about each school.

If you still can't find the school right for you, trust your heart to make your decision. You have been accepted into a medical school, there is no wrong choice from here on.

Finally, applying to schools and deciding which school to attend are two different things. You should always apply to the schools that you can see yourself attending. The more schools you apply to, the better your chances are for an acceptance letter. Deciding on the right one depends on which schools have accepted or rejected you.

Often times, the school you loved will reject your application or put you on a waitlist. If this happens consider another school further down on your list. In this case, I would suggest accepting their offer because "a bird in hand is worth two in the bush." You applied to medical school to become a doctor. You may not have gotten into the school you wanted, but you do have an opportunity to pursue the career you wanted and that is the bottom line. If you have one acceptance letter, or multiple acceptance letters, consider yourself lucky that a medical school has chosen you.

REFERENCES

1. "Canada's Best Schools: 2014 Maclean's University Rankings - Macleans.ca." *Macleans ca.* N.p., 31 Oct. 2013
2. *"2014 Maclean's University Rankings By Macleans | Theguidebook.ca." Theguidebookca RSS. N.p., n.d. Web.*

3. http://admissionmyths.med.ubc.ca/discover/is-medical-school-right-for-me/
4. www.macleans.ca/news/canada/how-much-they-pay-for-it/
5. "Report on Canadians Studying Medicine Abroad - CaRMS." *CaRMS*. N.p., 2010. Web
6. "Canadian Resident Matching Service - Service Canadien De Jumelage Des Résidents | CaRMS." *Canadian Resident Matching Service - Service Canadien De Jumelage Des Résidents | CaRMS*. N.p., n.d. Web
7. www.carms.ca
8. "CanMEDS: Better Standards, Better Physicians, Better Care." *Royal College*. N.p., n.d. Web

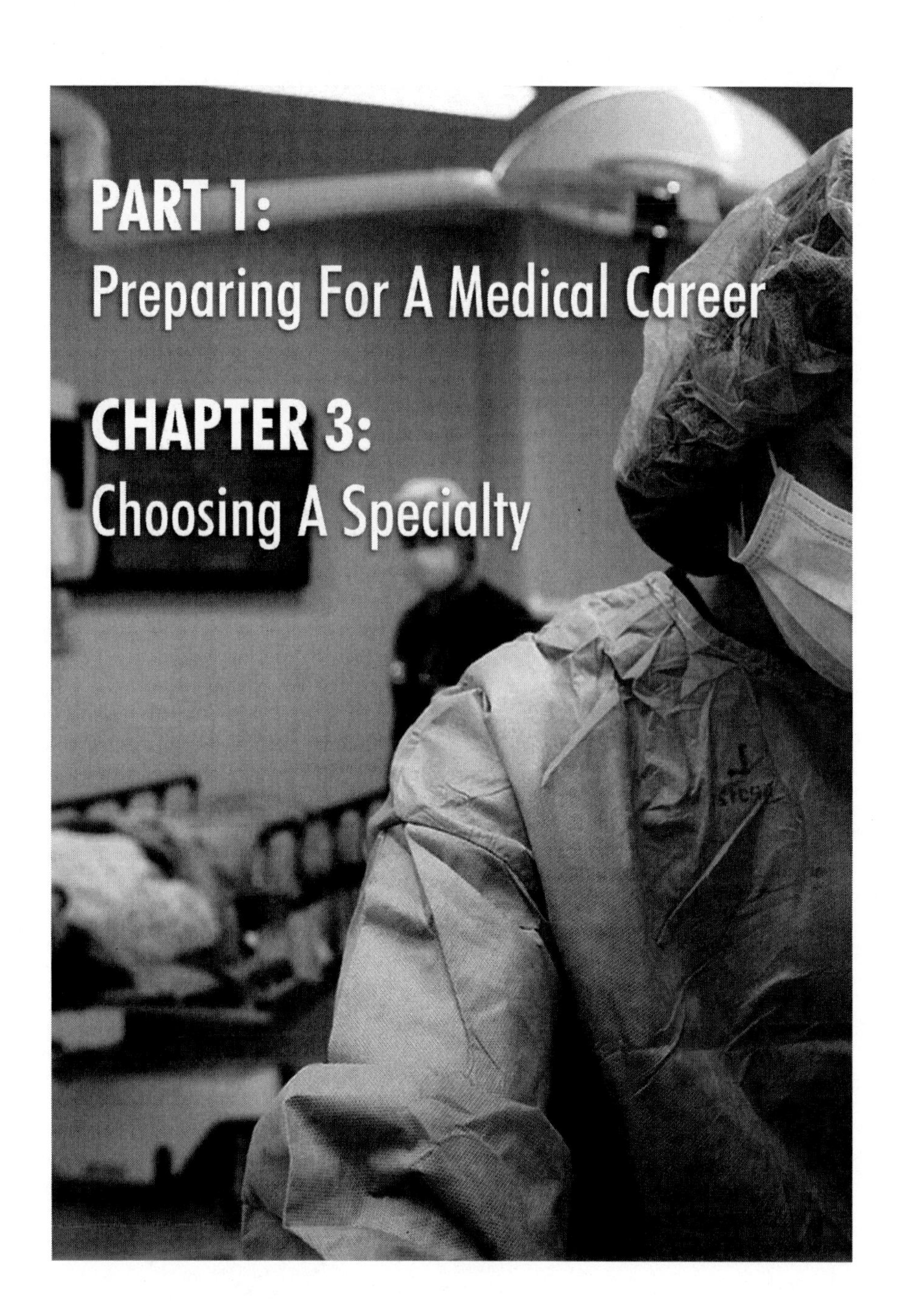

PART 1:
Preparing For A Medical Career

CHAPTER 3:
Choosing A Specialty

These are many reasons why medical students choose a specialty. Most medical students decide what specialty to pursue when they are in their mid to late 20s, in their 3th year of Medical School. While many think they are informed by that time, in reality most of the students have not been in contact with clinical work yet, let alone subspecialties and they don't fully consider the "big picture" when deciding what they are going to do for the remainder of their careers. In all fairness, medical students are busy studying and do not really have the tools yet to decide what subspecialty they may like to do in the future.

When deciding on a specialty, base your decision on your interests, skills and personality. Not on other factors such as money, work hours, opportunities, likelihood of employment or unemployment at the end of your training. Here are some approaches to think about when deciding on a specialty.

Family Practice or Subspecialty

Let's start by narrowing it to a simple enough question that usually translates to people dividing their thoughts into family practice versus subspecialty and medicine versus surgery. In many ways that is an accurate way to think about it but given the complexity and overlap of many specialties, you can do many surgical things in fields that require more medical based training and vice versa. Examples of fields with medicine training and lots of hands on procedures include interventional radiology, emergency medicine, gastroenterology, cardiology and anesthesia. If procedures scare you, there are many fields that give you non-invasive treatment options such as: internal medicine, oncology, radiology, nephrology, genetic, etc. If you want to be a surgeon but not spend all your time in the OR, many surgical subspecialties have clinics. Another option is that you can also work in the ICU.

"Colin has completed his cardiovascular training and has done one extra year training in Cardiac ICU. He has found a position where he works half time in a cardiac surgical group and half time in the Cardiac ICU caring for the group patients. It suits his interests and he does not have to spend every day in the OR."

Along the path to becoming a physician involves many decisions that will ultimately determine the populations you serve, the type of work you do and the relationships you build. Deciding on a medical specialty is considerably difficult, most medical students find that there are many sub-fields they find engaging. If you don't know what kind of work a field does, get out there and shadow for a week. Although I understand you may have limited time with classes and studying, this will give you a chance to see the variation in a job and what you may like. If you're looking into a highly competitive field, do this during the first two years of medical school instead of waiting for your clerkship. If you like a certain field, there may be some preparation you need to do first and contacts to establish.

Determining Factors

Some will find that the flexibility of their personality allows them to enjoy various kinds of work. And still, others will find that they do not like doing patients/bedside care as much and that's ok too. There are many ways to contribute to medicine post-medical school.

Most medical students have loans and therefore may have a practical reason for a high-paying specialty choice. However, as you have likely heard before, money alone won't make you happy. You will be practicing your specialty for the rest of your life and even if you decide to become a plastic surgeon because of the big bucks often associated with the practice, if you don't enjoy the types of procedures and patients it entails, you might be miserable despite your big bank account.

"Scott was eager to have a big home, take fancy vacations and generally live a life of luxury. After doing his research and seeing all of the glossy ads in medical magazines for cosmetic surgeons, he realized this would be a great choice to reach his goals. He shadowed a private practice plastic surgeon who had a thriving cosmetics practice and performed many cosmetic surgeries, Botox, and other "beauty enhancements." After shadowing this doctor for a few days, Scott decided he would not enjoy caring for this patient population. However, he reasoned that he could 'deal' with it if he was making a lot of money since he could spend his leisure time as he wished. However, after considering that most of his waking hours would be spent in the operating room however nice the office and the OR would be, he realized that perhaps he should consider another specialty."

Many medical school students are, by nature, very competitive and if you want to join a high profile specialty you must enjoy the work you do first. Also, as you mature, your values may change. In your late 20s, spending long hours in the hospital may seem glamorous and appealing, but as you get older and have family and other responsibilities you may not want to work as intensely as some specialties may demand. Also, keep in mind that these high profile specialties such as neurosurgery, interventional cardiology, cardiac surgery, trauma surgery and oncology often require a tremendous amount of emotional stamina, which may affect your personal life. Do not underestimate how tough that work can be. You may choose a specialty that has 'easy hours', such as emergency medicine (EM), radiology, ophthalmology, anesthesiology and dermatology. The hours related to these specialties often are not what you may think.

Anesthesiologists routinely wake up early because operating rooms open early. Even though full-time emergency physicians put in about 35 – 40 hours per week, they work odd hours: evenings, nights, and weekends. Odd hours can take a toll in the long run, something that's difficult to understand when you are young.

Conversely, during a surgical rotation, you are impressed by the surgeon's attitude and personality. She represents the type of physician you want to be in the future. She is swift in the OR and deals with unexpected complications with control and calm. Yet she is also kind and compassionate for her patients and their outcomes. She is also a real team player who treats everyone on her team with warmth and support. You want to be like her when you grow up so you decide to meet with her to discuss the idea of becoming a surgeon.

At that meeting, she tells you that working in academic medicine has many demands. She must publish, participate in hospital committees, teach, do research, and attend grand rounds even when not presenting. You tell her that all you hope to do is practice community surgery so she suggests you gain exposure in the field of surgery "in the community."

During your winter break, you shadow a community surgeon. The work doesn't seem nearly as exciting as the work in an academic setting. The surgeon has busy, but lonely, days filled with OR time, outpatient visits and administrative work. So you think you may need to reconsider your choice.

During medical school, most of the people you meet and your clinical rotations will take place in academic hospital settings. Yet the majority of medical school graduates will not practice in these arenas; most will practice in community settings. The reality is that specialties are practiced very differently in different settings and many students select a specialty based on their understanding of how it is practiced only in an academic medical setting. You need to consider this aspect of medical practice; context is key.

So how should you decide on a field to pursue? Whatever your reasons for choosing a specialty, you need to fundamentally enjoy the subject matter, the disease processes, the type of practice and the patients for whom you will be caring. Ideally, you also want a career that will have longevity.

Think Ahead

In making a decision, it is essential that you view your life in the future. Try to fast-forward 20 years. It is not easy, I agree. Where do you want to be? How do you hope to be practicing? Find role models and ask them what they like or don't like about their specialties.

Would they make a different choice now that they have a more mature perspective? Many people who practice primary care have great lifestyles and can also practice for a long time because their practice may not be as physically or emotionally rigorous, they are more in control of their practice life, their work hours and their workload. There is also less administrative work, research and publications requirements as there is in the academic world.

Do your research, explore how your desired specialty is practiced in many settings and, most of all, and be honest with yourself. Medicine is not a solo sport. Consider who will be your colleagues outside of your field. For example, anesthesiologists have to work with surgeons, nurses, techs in the operating room but they can also work with internists in the ICU. Consider which kinds of people you have to consult or work side by side with and decide if those interactions excite you.

Doctors are talking about money and lifestyle. And too many doctors over the age of 50 are complaining that the younger generations are not committed like they are. They're foolish and so are you, if you don't take the lifestyle and family into account. What you want in the next 5 years in terms of your life outside of medicine (and yes you should have a life, happy doctors tend to do a better job) may be different 15 years down the line. Plan in advance. This doesn't mean you have to know for sure what you want but make room for friends, partners, children, parents or siblings in your plan. Or maybe you would like to take a vacation or fit in regular sleep. Times are changing and while your predecessors like me, spent all of our time in the hospital, you do not have to and you do not need to.

This may not matter to many of you, as it's hard to see now. There are still some technologies under development or yet to be discovered, and this will impact your specialty. But there are certainly some fields that are incorporating technology at an unprecedented pace to decrease the risk of our medical interventions, improve patient safety and working conditions such as interventional cardiology, orthopedic, neurosurgery, radiology and more. The impact of technology cannot be ignored. Look into how your fields of interest are changing and if you like the direction it is headed.

Stuck

So what happens at the end of the road if you haven't decided and are still torn between more than one specialty? Take time off. Don't take time off to do research you don't care about. Take time off to explore the fields you are considering. There are ways to make that productive, I am sure you can come up with ideas to work in the community, to shadow somebody of your choice, to do volunteerism and even developing international health interest.

Do not enter in a match. You're still an MD, just not a practicing one yet. You can enter the match later. Do something good in the mean time. Or you can enter a match in the specialty you want to choose at the time. You'll be surprised how sometimes we can fall in love with something. If that's not the case and you make the wrong decision, you can always change your specialty and do something else. Many will say it is difficult because of the match program bureaucracy, but it is possible. All you lose in the process is a little bit of time. But look at it another way, all you gain is time to confirm that something is absolutely not for you and you can go down another path. But, please don't stay in a field you don't like for any reasons. Many people change fields. You, your colleagues and your patients will be better off to have an engaged doctor that wants to be there.

Factors Affecting Employment

In 2010, several of Canada's national medical specialty societies reported to the Royal College that a growing number of specialist physicians were unemployed or under-employed. The Royal College undertook research to look into this highly complex problem, seeking to determine if unemployment and under-employment are the simple and inevitable byproducts of an oversupply of physicians — or if other, more subtle factors come into play.

The Royal College has unique access to specialist physicians in Canada and, as such, was able to collect a unique set of data to inform this report. The Royal College surveyed newly certified specialists and subspecialists online for this study and interviewed more than 50 persons with first-hand, in-depth knowledge of the issues. This provided important new insights.

Data from the Royal College's 2012 employment surveys (*"Mobilizing National Efforts to Improve Health Workforce Planning." Royal College. 2012)*reveal that employment issues extend across multiple medical specialties. Among those who responded to the surveys of new specialists and subspecialists, 208 (16%) reported being unable to secure employment, compared to 7.1% of all Canadians workers as of August 2013. Comparison by year shows that employment challenges increased in 2012 over 2011. Those employment issues increased by four percentage points (from 13% to 17%) for specialists and by six percentage points (from 15% to 21%) for subspecialists, the findings also show potential provincial variances, which merit further investigation.

Almost 22% of new graduates reported they are staying employed by combining multiple locum positions or various part-time positions.

The survey revealed that a substantial proportion of new physicians experiencing employment issues were from surgical and resource-intensive disciplines, including but not limited to: critical care, gastroenterology, general surgery, hematology, medical microbiology, neurosurgery, nuclear medicine, ophthalmology, radiation oncology and urology.

The economy is the main factor driving new medical and surgical specialist un- and under-unemployment in Canada. More physicians are competing for fewer resources. Many of medical and surgical specialties depend on health care facilities such as hospitals and their resources, including operating rooms and hospitals beds. Access to these resources directly impacts the physician's employment. While the health care needs of patients increase and the number of physicians and surgeons continues to grow, hospital-funding growth does not appear to maintain the pace. To control costs, resources such as operating room time are cut. Physicians are competing for fewer resources., plain and simple.

Another factor is that the stock market's relatively weak performance in recent years means many medical specialists are delaying their retirements. This means long-held positions are not freeing up as expected. Inter-professional collaborative practice in which multiple health workers from different professional backgrounds provide care to patients is increasing in Canadian healthcare. With this development comes a new range of health professional roles and responsibilities that affect how Canadians interact with medical services and physicians.

These new roles associated with inter-professional care models complement and in some cases substitute physician services, making it possible to increase the amount of specialty medical care that physicians and surgeons provide without increasing the number of jobs.

Some teaching facilities are adapting models of care, which includes health professionals that complement or substitute for physicians.

Approaches for allocating residency positions can vary greatly among the provinces and territories and decisions are broadly based on scant information about longer-term societal health needs and health care resources. In addition, new medical graduates may not remain in the jurisdiction where they trained, creating further imbalances in medically trained professionals.

Most health professions develop a personal culture of practice, which is motivated by both altruism and self-interest. Given that the availability of clinical and practice resources is generally fixed (this is especially true of operating room time) established specialists can be reluctant to share resources. This is because they wish to protect their access to clinical resources for their patients and their incomes. Income protection has become especially important for those whose retirement portfolios were negatively affected by the recent economic downturn.

The importance of career counseling for specialists cannot be overemphasized; many have not received any career counseling, and it is an issue. It is certainly a missing link the medical training curriculum as a whole. Lack of transparency about available jobs has also hindered the ability to find suitable work. And finally, personal factors will always influence the choices that physicians may make about the type and location of practice they will pursue. Newly published data shows that the number of nurse practitioners has more than doubled over five years, including a 25-per-cent jump in a single year *("Nursing Leadership." The Primary Healthcare Nurse Practitioner Role in Canada, 2010).*

At the end of 2010, of the 2,486 nurse practitioners in Canada, 1,482 work in Ontario. Quebec, the next most populous province, has only 64 NPs, while B.C. has 129 and Alberta 263. Yukon is the only jurisdiction that does not license NPs. Policy makers believe NPs can improve access and reduce costs by doing work that was once the sole purview of physicians. NPs are especially popular in rural and northern regions where there is a shortage of family doctors, with patients who suffer a multitude of chronic illnesses and are ill suited for a traditional seven-minute doctor's appointment. In hospitals, NPs work principally in emergency departments, where they do triage and handle less acute cases, or they oversee clinics that treat patients with chronic illnesses such as diabetes, provide dialysis or cardiac rehabilitation.

New specialists who lack access to suitable hospital resources, or who cannot find jobs in the settings they require, are developing 'tailored' or 'morphed' practices that allow them to work within the resources available to them. These new practices do not embrace the full spectrum of specialists' abilities, resulting in skill loss and under-employment (such as when a surgeon cannot operate). This situation is called "brain waste."

Just under 20% of recently certified medical and surgical specialists have not found positions. These same specialists said they would look for work outside of Canada. The predicted physician shortages in the US may become a market for Canada's unemployed specialists, prompting a "brain drain." The fix is to invest wisely in the system, as well as provide incentives to those areas where there's a higher societal need. Areas like primary care, mental health, palliative care and geriatrics are examples of this. Moving to meet these trends is a powerful insurance measure.

The problems can also be traced back to medical schools, where there is scant science behind deciding how many positions to allot to each field. Determining optimal numbers of training positions is not easy, but the practice of doing so is definitely lacking. In response to the outcry over long wait lists, provinces have in recent years significantly boosted medical school enrollment and the number of on-the-job training positions: two-year family-medicine residencies and five-year residencies in a specialty. Even if making changes to the admissions quota were a good thing, it would take a minimum of six years for family physicians and anywhere from seven to nine years for specialists to graduate. So, we're looking at roughly a decade before we start to see the effect of any changes we make to the medical school admission process. And it would be at least half a decade before we see the effects of changing specialty-training quotas.

It has become clear that medical workforce planning must consider the availability of practice resources as well as the health needs of the population. This approach would help physicians do the work they have been trained to do for the future.

Military Career

Few people would consider pursuing a military career unless their medical education was going to be funded by either the Canadian Forces Health Services or the Health Professionals Scholarship Program of the US Military Medical School in Bethesda.

The Canadian Armed Forces Health Services will pay recruits to complete a medical education program at a Canadian University. Their program covers the cost of tuition fees and educational expenses (including books, instruments supplies, student fees and registration costs).

For the duration of your studies and residency you will receive a full-time salary including medical and dental coverage, as well as vacation pay, in exchange for working for the Canadian Armed Forces. The length of time that you will be working for the Canadian Armed Forces will be determined by them. They will take into account the cost of your medical education and then provide you with the number of work hours required.

In summary, choose the path that you think you will like. It sounds simple, but don't think that makes it any less true. Unfortunately, the decision for the match has to be done early, which can be unsettling for many of you because of understandable uncertainty.

Know who you are and be realistic about your strengths and weaknesses. Learn more about the subspecialty you would like to choose. Try to shadow a mentor if needed be. Remember that you have not committed yourself for life in the choice you made and you can always move to another subspecialty of your choice, albeit, it will cost you some time. And as far as the unemployed physicians rate of 16%, it is in the context of today's health care system, and it may be different by the time you have completed your training. Therefore, I would say, choose what you like and if you can, choose what you would love to do.

REFERENCES

1. "Mobilizing National Efforts to Improve Health Workforce Planning." *Royal College*. N.p., 2012. Web
2. "Nursing Leadership." *The Primary Healthcare Nurse Practitioner Role in Canada*. N.p., n.d., 2010. Web.

PART 1:
Preparing For A Medical Career

CHAPTER 4:
In-Training Tips

"Being a Student is a Full Time Job"

After 4 years of medical school, you graduate with a degree in medicine. This is a great achievement. In the past, people had to add a one-year internship after 4 years of medical school to graduate with a degree in medicine.

There are two qualifying exams: LMCC (License of Medical Council of Canada) Part One which happens at the end of medical school. This is required in order to enter postgraduate training. It is a daylong exam, the first part being multiple-choice questions and the second half being "clinical decision making" short answers. LMCC Part II happens after at least 12 months of postgraduate training and is a series of clinical stations and is required for independent practice, and more importantly with the College of Family Medicine.

For historical purposes, many years ago we had to do one year of an internship after 4 years of medical school, and then at the end of the internship, write the medical doctor/family practice Royal College Examination. You would then receive your medical license to practice. There was a choice of surgery, family practice or to enroll in a specialty. As a surgical specialty, a person had to complete 4 years of general surgery, pass the Royal College General Surgery certificate before doing a sub-specialty like cardiovascular and thoracic surgery, neurosurgery, urology, etc. As a medical specialty, a person had to do 3 to 4 years of Internal Medicine, pass the Royal College Internal Medicine Examination and continue with a subspecialty like cardiology, nephrology, etc.

The Royal College of Physicians and Surgeons of Canada has reviewed the residency-training curriculum several times over the years. The current format includes 4 years of medical school with a medical degree, and then 2 possibilities: a person can choose family practice and then enroll in a two-year residency before getting a license to practice, or a person can choose a subspecialty tract.

The specialty tract has combined Internal Medicine in the 5 to 6 year medical specialty program and included general surgery PGY 1-2 with 3 to 4 years of surgical subspecialty. For complete information on training requirements for the Royal College of Physicians and Surgeons of Canada accredited residency programs, please consult their website *("Accredited Residency Programs (ARPs)." Royal College)*

With the latest subspecialty residency programs, the greatest risk is at the end of the residency when there will be the final and only Royal College of Physicians and Surgeons of Canada Examinations (written and oral). There are in-training examinations/tests that need to be completed to qualify for the final examination.

When you have passed the exams, you will then receive your subspecialty license that will allow you to practice in that specialty only. The greatest risk comes from the stress to pass the Royal College Examination because if failed, one has to complete a full year of training before re-presenting at the exams.

The advantages of the latest Royal College Residency Programs format is that it allows for a longer more complete training. You will be exposed to a subspecialty, and today every subspecialty is developing at a fast pace with new science and technology.

As you start your first rotation, the feeling of achievement for being a medical doctor may be superseded by a feeling of anxiety. You may also feel inadequacy, or be overwhelmed when you think about the amount of material there is to understand and the skills required for mastering new technology, surgical or medical. In the beginning of the residency and until you develop the level of comfort and knowledge needed, it may feel stressful but it will become easier.

Being a subspecialty medical doctor-in-training and being a resident, all of this is a new level of study for the next 4 to 6 years (and perhaps more, if you include a fellowship).

It is a "full time job" you have chosen to prepare for your career. It is very unlikely you will lose your "job" if you follow the steps of the residency program (despite a good or bad economy). You work in a safe, clean, peaceful and professional environment, which will provide you with learning opportunities. You will also have a chance to build friendships and professional relations with your colleagues.

An important point to remember if you have challenges during this period: you are being paid to learn and for the work you do in the hospital. During your 4 to 6 years of training, the residency salary compensation will go up incrementally (Table 1) *("Salaries and Benefits - CaRMS.")*.

Table 1

Salary Scale Ontario Residents

Position/classification	Effective 7/1/2012 Ontario	Effective 7/1/2012 B.C.
PGY 1	$51,065	$49,934
PGY 2	$59,608	$55,705
PGY 3	$63,230	$60,702
PGY 4	$67,512	$65,341
PGY 5	$71,995	$70,268
PGY 6	$76,210	$75,022
PGY 7	$79,220	$79,951

If you are paid for hospital work, learning opportunities or both, you may need to put things in perspective and recognize that in most professional university programs, students start to get paid only when they have a full time job (accountant, engineer, lawyer, nurses, teachers, etc.).

As the residency training salary compensation increases every year, you quickly reach and surpass the Canadian median salary of $46,085.

Some may not like the idea of being on-call but this is one of the requirements of the residency program. This will be a part of your medical practice in the future. Enjoy your residency training time, as it will provide you with the opportunity to learn and expand your medical practice.

Rotations

When you have completed 4 years of medical school, you will have a vast but mainly generic medical background.

Do not be mistaken: it is important material to know. In your future training and practice, chemistry, biology, physics, anatomy, genetics, Chemotherapy, cardiac drugs, computer-aided surgical technology, biostatistics and much more…all of this knowledge will be useful regardless of what you are learning. As you start rotations at the beginning of your subspecialty training, you may feel that some of what you are learning is irrelevant toward your career as oncologist, family doctor or orthopedic surgeon. Why would an orthopedic surgeon or cardiac surgeon need to know about gallbladder surgery? All opportunities to learn about medical and surgical disease are important as well as learning general surgery. Learning how to cut, to sew, to handle surgical instruments and wound infections, etc., are important. The same applies in a medical subspecialty. Medical or surgical, the rotations of your first and second year of training will allow you to acquire what is referred to as medical maturity, medical judgment toward decision-making and medical preparedness toward your real field of interest.

Maturity, judgment, surgical skills, technology skills such as cardiac catheterization and intervention, of any scope, all take time to acquire and the beginning of your subspecialty training is as important as the latter part, which obviously is more challenging and more specific to your interests.

"Kim was in her second year of general pediatric specialty and had a rotation in a small peripheric hospital with a small general pediatric clinic and ward. That rotation was based in providing exposure to a general pediatric practice. Unfortunately for Kim, there was very little teaching and many days, she thought she was irrelevant."

Rotations that are boring, that do not provide the appropriate learning opportunities, and do not meet the Royal College teaching and curriculum goals, do exist and will continue to exist. Not every single rotation will be perfect, meeting all the teaching goals and providing all the essential learning opportunities. You do not need to get frustrated by this. If you face this situation, have a constructive discussion with your program director during the rotation or after to explain the issues and provide positive recommendations.

Secondly, do not think you are wasting your time. Bring your books, your tablets, your laptop to work and catch up on your readings, your research if you have any projects, or your writing. All these activities are useful and need to be done at some point, so you can use this time to get ahead. Do not lose time waiting for teaching opportunities: create your own and you can make the most of this time. Turn it into a positive experience. This time will pay off in the future.

Workings Hours and On-Call Time

It is difficult to find a balance between working hours, on-call hours and learning opportunities. The Royal College of Physicians and Surgeons of Canada and the Accreditation Council for Graduate Medical Education in the United States have both had multiple committee reviews of optimal work hours during a residency training workweek. Shorter hours make residents more alert and better able to learn, but also results in less exposure to patients and the course of their illnesses. Unfortunately, there is no scientific answer to resolve the ratio of optimal hours for education and rest. Research has not been able to establish an exact number of hours per week below, which residents may safely and effectively learn and participate in patient care.

Adding to this difficult issue is the fact that the residents are vital to the provision of patient care in a teaching health care system. This system faces a serious work force shortage. Some residents may feel it is not their responsibility to cover for the hospital or the physicians. It is a trainee's duty to provide exceptional patient care coverage through an on-call schedule in exchange for learning opportunities.

Over the last 40 years, there have been many changes in the training curriculum. A policy for the number of work hours and of on-call hours per week has been implemented. Not only has the number of hours on-call been limited to 24 hours, there are now talks to decrease the on-call hours to 18.

A trainee cannot be on-call more than once every 5 days and in many subspecialties (anesthesia, ICU, ER, to name a few), the residents are off the day post-call. The maximum of hours worked per week is now down to 50 hours. All these changes have had a positive effect and allowed for a better learning disposition with better studying capacity.

During the 2 years of Family Practice residency and the 4 to 6 years of subspecialty residency, there will be many occasions when a new patient, a complex case, a longer than expected or delayed operation or a last minute meeting will change the order of your work day. You will be challenged to adapt to sudden changes, to face the decision of staying late or going home, and potentially losing an opportunity or capturing one.

Medicine is not a 9 to 5 job and many more changes during workdays will happen in your career. For a surgical resident or any medical resident in a subspecialty requiring technological skills (cardiac catheterization and intervention, catheter intervention, any types of scopes, ultrasound, arterial line, central line, etc.), come prepared to the room where the procedure will take place. Know the patient, the previous lab tests and results. Show interest in your patient, not just the test. This will be an excellent habit for your practice.

You should operate and learn about the technology you need to master, then repeat over and over again whenever possible. You can never be too proficient. Do not avoid long difficult cases. Work with every single one of the attending physicians to learn different views and different approaches. Do not limit your area of focus too early in your residency. Any extra days you take off are lost opportunities, remember that.

Repetition is the key to unparalleled surgical skill, technical skills, low complication rates and a broader repertoire of knowledge. The more you do as a resident, the more you can offer your patients in the future and the less you will need to refer to others.

"Peter is in his fourth year of anesthesia residency. During his pediatric anesthetic rotation, he was assigned one morning to the cardiac room. An emergency second open heart case started late, at 3 p.m. Around 5.30 p.m., Peter discussed with the anesthesia attending that he was going to go home, as it was the end of the day. In doing so, he missed the opportunity to learn about the treatment of severe pulmonary hemorrhage when the baby was weaned off cardiopulmonary bypass. Peter's reaction the next day was that he would not do pediatric anesthesia."

Leaving late is not an every day thing. It is a question of judgment, balance between a learning opportunity, time spent at work, working long hours and arriving home late. You will learn something new or acquire a new skill when you put more time and effort in. You will most likely not regret the time spent.

One last suggestion relates to the organization/planning of your day off, post call. There is certainly time needed for rest and perhaps you worked all night. But it is a good principle to oblige yourself to do 2-4 hours of work: reading medical books or articles, writing a manuscript or preparing a lecture. Include such learning activities with your sleep and exercise schedule in your post call day. Take the approach that it is not a day off.

In fact it is a good piece of advice to read at least one hour per day during training. It will slowly prepare you for the Royal College Exams. Your knowledge base is critical for the proper care and diagnosis of your patient, much more than the simple necessity of your exams. Someday you will be on your own listening to a patient giving you a long list of complaints/symptoms, expecting an immediate diagnosis and treatment. You can leave the room to consult another colleague as will be needed with very difficult and complex cases, but it should not become a habit. Read as much as you can and as often as you can now, so you are able to provide your patients with the best care. You should maintain this habit during your practice, to keep up with the latest information, treatments and technology.

Communication

As a doctor, we do not get courses and education on the art of communication and yet a large part of our work is communication with patients, staff or, colleagues, in a team-working environment or on a personal one-to-one basis. You will encounter a spectrum of attitudes and behaviors, both good and bad.

During my residency, I had three major aggressive interactions with another resident. These interactions upset me very much. The only thing I was thinking at the time was standing my ground, but I could see it differently today.

When I started my practice, it did not occur to me that my medical colleagues could have an unprofessional and un-collegial behavior towards each other. Even more so, my senior mentor was a master at public speaking, of being controlled, mature and diffusing difficult situations. I learned from the right person. As I grew older in my practice, my thoughts on professionalism and collegiality among physicians changed due to the reality of my life experiences.

During your career you will probably encounter conflicts and unprofessional behavior. In any situation that arises, a simple approach is to take a deep breath, control your emotions, ask questions to help to understand the issue and formulate answers or comments. Sometimes it is better to walk away and regroup. It is also easier said than done.

Do Not Rush

Medical school provides a very broad medical framework, which, as stated earlier is certainly not irrelevant. Many aspects of this information will be used during your training and your professional career. The same may apply at the beginning of your subspecialty training, meaning the beginning of the residency will be more generic with increasing focus as you progress.

For instance, in most of the surgical subspecialties, the first 2 years (PGY 1-2), will be mainly general surgery. A neurosurgeon or cardiac surgeon or ENT surgeon will not do gall bladder surgery, ever, but the training is useful in providing learning opportunities of surgical principles, of developing decision-making judgment, of surgical skills and acquiring surgical maturity. The same applies in internal medicine and subspecialties, as far as acquiring decision-making, judgment and even technical skills. As your interest becomes more and more on the core focus of your career, it is still important to devote as much attention as possible to these rotations.

Most of all go to work happy every day, concerned about your patients and your learning. Do not rush through residency or a fellowship focused solely on just getting the job done. By only going through the motions you are not helping yourself and can establish a pattern where you tend to cut corners on procedures. Do not be in a rush.

As you get closer to the last year of your training, the appeal of a full time position, of your own private practice, will become more intense. The prospect of a sizeable income will become more appealing. The recommendation is again; do not rush, for several reasons.

The most important reason is that the time you are spending during this period learning new skills and being exposed to diverse cases is a once-in-a-lifetime experience.

With the demands of a busy practice, a busy office, or family responsibilities, it will be difficult to attend refresher courses or new technology training courses as often you would like; attending regular national and international conferences may also be a challenge. Therefore, the more you learn in training, the more proficient you will be, despite always having the possibility in your practice or working environment to consult colleagues for particularly difficult complex cases. You want to feel secure in your decision-making and also gain the respect of your colleagues.

Life Issues Considerations

Life is a balance and a choice. Being in training and on-call can keep you busy. You can achieve balance by eating well, doing regular exercise, and making sure you get enough sleep. When we are young, we believe we can handle an intense schedule and long work hours. To some extent this is true. But if you establish good, healthy lifestyle habits now, it will be easier to incorporate them later in life and thrive in the years to come. As you age, you will become busier with your practice and your family life. Establishing these healthy balanced lifestyle habits now will help you live a healthier, happier life for the long term.

As you get older, it gets harder to establish a life balance. You will get very busy and perhaps gain weight. With the long workdays, you will think that you do not have time to go to the gym and keep fit. In fact, it is possible and easier than you think to maintain these good healthy lifestyle habits.

Before you can look for a job, you must consider the type of practice you want to be in: academic or private practice, be it fee-for-service or an alternate payment plan. It is often difficult to decide on a career in academia versus private practice, but with each rotation you can determine what you like or don't like and will establish if you like teaching, writing manuscripts, or just focus on clinical work.

Once you have established what you career path should be, start looking early in your training. This is recommended before your 4th year in subspecialty and as soon as you start your family practice residency. As you will have less time to find a family practice place, look at the area you are interested in, and at hospitals you are interested to have an academic career in. Find out the physicians in family practice or in your subspecialty who work in the area you may be interested in. You will get an idea of the available expertise, competition and potential all at the same time.

Send an email to the physicians you find are an appropriate fit for you. Introduce yourself. If you are really committed to this area, contact the relevant people and meet with them to introduce yourself and assess the possibility. Be persistent and stay in contact. Networking is very important mainly because you will have only a few years of training. During this period, physicians that you are speaking with will decide their recruitment plans and hopefully present you with a job offer in the future.

You will be receiving an incremental salary over 2 to 6 years. As you are a part of a residency-training program, you will still qualify as a student and therefore if you have student loans from college or medical school, you do not need to start paying them until your residency training is over.

But over the next few years of your residency training program, you should, during the first year, develop a financial plan for the time you are in a residency program and until you start your practice. A salary of $30,000 to $72,000 is still a fair amount of money. The most important rule is not to get in debt, or not to increase the debt you may already have. The exception would be to buy a condo/house, as it is a good investment whether you plan to stay in town or not.

If you are single, planning to get married, or married, the amount of disposable money might even be more as your spouse may be working. After establishing a small financial plan, you should consider 4 points:

Start to pay your student loans down even if it is by a small amount. Theoretically you do not need to pay it now. However, in the future, there will be expenses to establish your practice and it could be helpful to have the extra borrowing room.

1. Start to put a small amount of money toward your RRSP. The tax benefits will be helpful.
2. Open a savings account and put money aside slowly to put it towards paying for your Royal College exams and the travel expenses to attend exams. The costs can be easily over $5,000

to $6,000. By having the money upfront, you will not need to borrow.

3. Do not borrow against your future earnings, under any consideration. Every doctor in-training knows that when they start in practice, there will be a major income increase and which will increase with the development of your practice over years. Your lifestyle will change, but you do not need to get into more debt than you already have by buying cars, travel and furniture, etc. You can live within your means even if you have to support a spouse or a child. It is good if you can adhere to your budget consistently. This will put you financially ahead early in your practice. Remember it is easier to spend money than make it.

When the practice is established, there are really very few ways for a doctor to increase their income: working harder and seeing more patients, working more hours and doing more calls.

All of these options come at the expense of your health and your personal and family life. We all have a mortgage and some loans, but accumulating debt does not create a sound financial proposition. Use your years in training as an exercise to establish your long-term financial plans.

REFERENCES

WEBSITES:

1. www.royalcollege.ca/portal/page/portal/rc/common/documnent
2. www.royalcollege.ca/portal/page/portal/rc/credentials/start/exams/candidate_information/exam_registration/cxam_fees
3. "Salaries and Benefits - CaRMS." *CaRMS*. N.p., n.d. Web
4. www.cga_ontario.org/assets/file/taxtipsforstudents_2011_pdf

PART 2:
After Graduating

CHAPTER 5:
Looking for Work Opportunities

Preparing for Your Job Hunt

As you are completing your training, you will need to start the process of the job search. A residency-training program will not assist you with your job search. Some residents may think so, but sadly it is not included as an obligation of a residency-training program. As stated, the obligation of a residency-training program is to train you following the requirements and the competencies based curriculum of the Royal College of Canada.

Therefore, you will need to write a curriculum vitae (CV) or resume, to search for a job on the Internet, to talk with friends, staffs and mentors, to attend national and international meetings, to establish a list of potential areas you are interested in, to send letters to different programs and to get ready for interviews.

The best advice for those starting their job search is to start at least one year in advance. The best sources of information are around you: your friends, other residents, staffs and mentors you have worked with. They will have a network of physicians they know locally, nationally and internationally. You will often find that some of the best positions available are not widely advertised and you will hear about it from colleagues, mentors and friends.

Careful preparation of your CV is one of the most important steps in preparing for your career in medicine. The review of applicants' CV by any potential group practice/employer/colleague is the most important step in choosing a good colleague for the position. You have to be prepared. Being a doctor, being a specialist, does not provide a job automatically. Some fields have fewer jobs available than others and it may be difficult to find a position that meets your goals.

In preparing a CV there are several steps to consider:

1. Never submit a handwritten CV; this is not professional enough in today's computer era.
2. Never submit a CV without a cover letter.
3. Never include your personal demographic, marital status, family, children, picture of yourself (passport picture).

4. Do include your education awards, diplomas, and publications.
5. Do not hire a firm to do your CV; it is expensive and the fancy technology of designing your CV will not necessarily help. I am sure you can find CV framework example on the Internet, or from your friends.
6. Do not embellish or exaggerate your information. It does not help.
7. Do not include awards, achievement activities before college. Even from college, include only the very specific important items such as a scholarship.
8. The total length of your CV should not be more than 6-8 pages. Interestingly enough, employers/colleagues may lose interest with a CV longer than this.

A concise, well-organized CV will impress those reviewing it. They will be able to recognize your organizational skills and your serious interest in the position being offered.

Your cover letter should include a statement indicating why you are interested and how the position fits your career goals. Also, a brief statement about the strong points in your CV and why you are a good candidate for the position. Restate the information about how to reach you and names of references.

Interview Process

Preparing for an interview is basically like going through an evaluation process. You should be prepared to articulate your strengths and weaknesses, including what you will add to the group if you do have a special interest or skill set. If you are nervous about the interview, practice with a colleague resident or with one of the staff. It will help you put your thoughts togcthcr.

Although you may think it is somewhat irrelevant, make sure you have an appropriate wardrobe. Most medical professionals expect interviewees to wear business type apparel. Certainly the apparel does not have to be expensive or of the latest style, but proper and clean.

Hairstyle is less important, but being groomed is appropriate. Before the interview, you should ensure you know all the information in your CV and you should have reviewed the position you are applying for. By understanding the position you are applying for, it will help you to better direct your questions in the interview. Arriving early is always a helpful step to familiarize yourself with the environment.

Generally, an initial interview is mainly for exchanging information, and getting to know each other. You should explore how you will fit into the position, the practice itself and how you would enjoy the particular work environment. When joining a small group, it is important to meet all the members of the practice who will partner with you.

Discussing specific agreements regarding salary, benefits, work schedule, should be done in the second interview. In many hospital/university/academic positions you apply for, you will probably be meeting a large number of people (specialty colleagues, other specialty colleagues you will be working with, managers, directors, department chief, and perhaps more).

It is a good idea to have a small notebook to write down their names and their respective roles of key personnel. It can get a bit confusing and overwhelming sometimes to meet so many people in a short period of time. It will also help you formulate relevant questions directed at the right person during the second interview, which is an underestimated upside. You will probably be asked to make a presentation on one of your research projects.

It is an important that your potential colleagues understand your interests, your research activities, and how you are able to present information and data in a public setting. Do not worry; no one will expect you to be a charismatic public speaker. If you happen to be good at public speaking that is a bonus.

At the second interview, you will need to prioritize the information you will discuss to cover all the specifics of the position. You will be meeting with fewer people, and most likely the people who decide the specifics of your employment. You will need to be prepared to address these concerns and questions:

Office space
Support staff
Salary or fee for service scheme
On-call schedules
Support for new equipment
Office hours
Clinic hours
OR time
Number of beds allocation
Hospital support
Research allocation (space, protected time and funding)
Benefits, moving expenses, vacation time
Maternity leave
Insurance, liability
Potential buy-out in the practice

Along with anything else you want to include. At the second interview, your spouse/partner should accompany you to tour the environment, community, housing, schools, and job opportunities for him/her if needed.

A lawyer should review your contract and ensure all the above information is included, that you are well protected, and that your employer/colleague and yourself have the same understanding of the employment conditions and the legal requirements of the position.

Income and Financial Arrangement

The Canadian system is vastly different than the American one. Provincial Medical Services (MSP) *("Canadian Health Care: Provincial Health Insurance.")* insures medically required services provided by physicians, health care practitioners, laboratory services and diagnostic procedures. It processes claims by these different groups of providers.

To be insured, the population of Canada needs to register to their respective province and pay their annual fees to the MSP. MSP processes billings and payments of physicians and practitioners on a fee-for-service and alternative payment basis.

Billing and payment data is used by provincial governments for planning, analysis, and management of their health care budgets. Over 80% of Canadian physicians are on a fee-for-service plan at provincial government level. There has been increasing number of hospital fixed-based salaried physicians with benefits, contract salary physicians, academic salary based physicians with benefits and physicians under sessional fees.

The fee-for-service payment plan is directed by each provincial health care budget, determined on a fee per medical act, and the amount of remuneration per act may differ from province to province. This fee-for-service payment plan is not related to productivity and/or performance, and remuneration is based only on the number of medical/surgical acts per physicians.

Increasingly, for physicians under remuneration plans other than a fee-for-service plan, the hospitals/universities/governments are pressured by budget restrictions to establish a system whereby doctors' productivity and performance can be evaluated.

The governments want to ensure that the public receives appropriate health care services for the fees they pay.

In the past, it was expected that doctors would provide excellent care, but there was no real measurement of productivity and performance.

With new alternative payment schemes (contract, sessional, salary with benefits, etc.), increasingly any new contracts will include productivity and performance assessments. Doctors will be obliged to keep data/information about their work schedule, their work productivity, research time allocation, publications, meetings attendance and holidays. Many of us may think we are punching a clock card, our professional autonomy is being taken away or that we are becoming bureaucrats, but it will be the new paradigm of medical practice.

Some family physicians and specialists will consider buying a practice or buying into a practice. One needs to be careful with this. The first question that arises when it comes the time to investigate the financial aspects of a practice or a group practice is what information and how much information is appropriate to ask? And what will you be entitled to obtain for assessment purposes? Many young physicians with little experience may feel shy and nervous about requesting sensitive financial information from a practice.

Most physicians selling a practice or hiring a colleague should view the request for information as appropriate and as a sign of your diligence and professionalism. Any unwarranted or excessive hostility or inability to produce basic information or documents should raise a red flag. The potential risks and rewards of this investment are too great to ignore.

Depending how financially savvy you may be, I would recommend that you enlist an accountant to help review the financial information, the balance sheet, the accounts receivables, the financial debt of the practice and, the liabilities and the risks.

As a final note, never sign a contract that does not fully address all the expectations and representations in the formal and final offer proposal. Often you may feel it is acceptable to leave some issues because they seem unimportant to you, or that your colleague promises it will be addressed soon or that you do not want to be difficult. But it could later create some unhappiness or even difficulties later on.

Let your lawyer handle the details even if the person hiring you does not think it is necessary. Each party will know exactly what has been offered and agreed upon, and each party will benefit of the rewards of a good working relationship.

REFERENCES

1. Holliman C.J.,
   Resident's guide to starting in medical practice
   Williams & Wilkins, Baltimore, Maryland, 1995
2. Shaw K.K., Raj J.K.,
   The Ultimate Guide to finding the right job after residency
   McGraw Hill, NewYork, 2006
3. "Canadian Health Care: Provincial Health Insurance." *Canadian Health Care: Provincial Health Insurance.* N.p., n.d. Web.
4. www.msp.canada
5. www.health.gov.ca/msp/

PART 2:
After Graduating

CHAPTER 6:
Establishing a Medical Practice

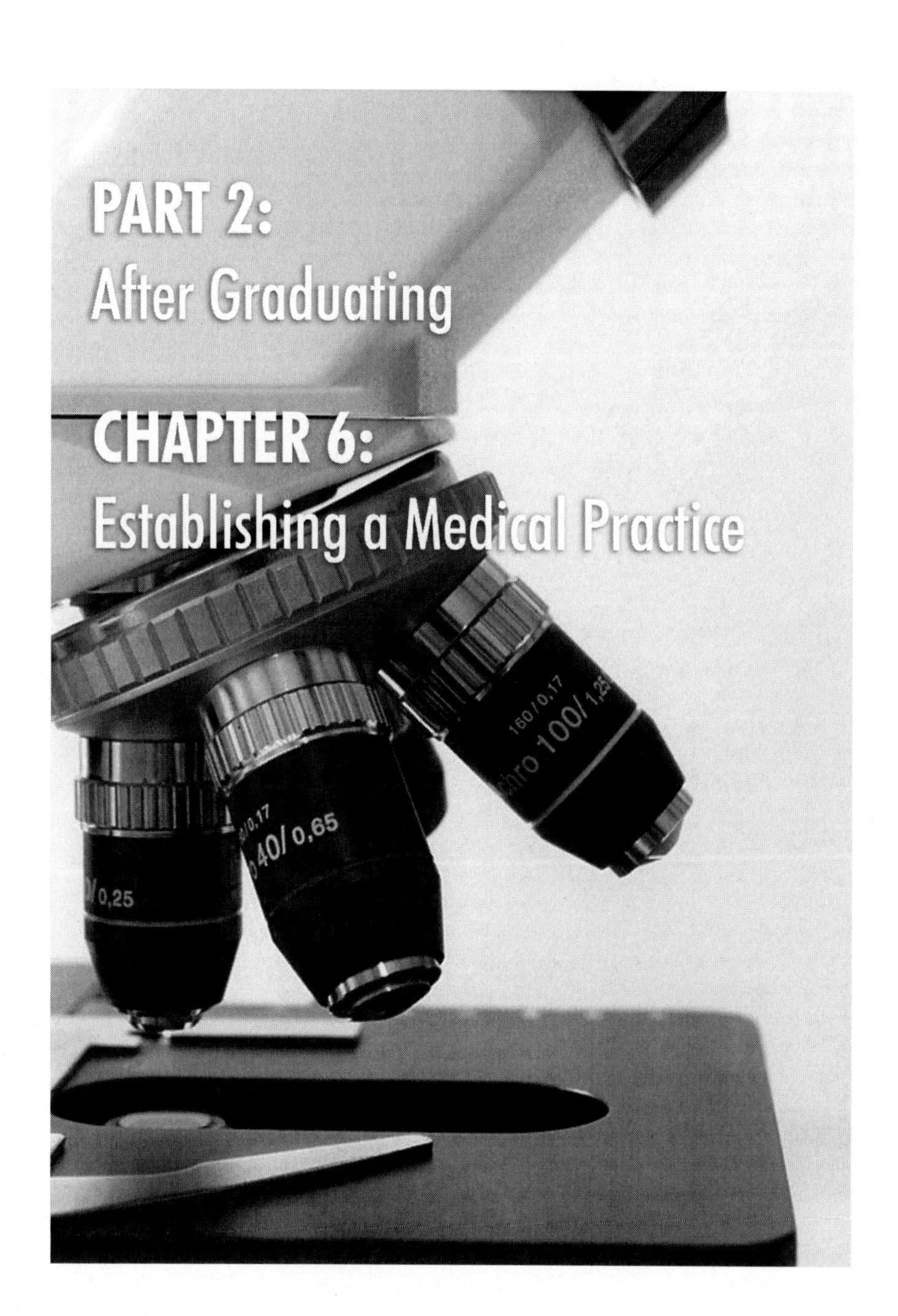

There are many different types of medical practices to join or start, each with its own particular pros and cons. Just as it is important to know yourself, it is also important you know what you are looking for in a practice. It is crucial to find an ideal location, to find a nice group of colleagues to work with, to get the pulse of the practice you are looking at sharing, developing or buying, and ultimately to find the right fit that will meet your career goals.

There are several types of practice setups. The main categories include the following:

1. Solo practitioner (starting your own practice)
2. Joining a solo practitioner
3. Joining a small group practice
4. Joining a multispecialty group practice
5. Joining a specialty group practice
6. Joining a hospital based practice
7. Choosing an academic practice
8. Choosing a military practice

Solo Practitioner

A solo practice can be a rewarding and enjoyable endeavor, but perhaps is one of the most difficult and challenging choices. You need to put careful thought into considering the challenging practice environment it entails. Obtaining financing to start a solo practice may be quite difficult as well. It could be a challenge to find a bank that will support your start-up with business loans and other products you may need. Information about your assets, income or potential income, collateral if you have any, this may be required by your banker to provide financing.

A solo physician will spend more time alone and has less interaction with colleagues than his colleagues working in a group practice. It is more difficult to discuss complex cases. Call coverage may also be difficult to organize.

One benefit in operating your own office is the freedom to organize your office the way you want and to hire who you want. Decisions can be made quickly and implemented without problems. Salary, vacation, and schedules can be easily controlled. Job security, scheduling, flexibility, the ability to complete tasks, the ability to provide your approach to medicine and your interpersonal relations with your staff. This is all under your control.

A solo practice can easily be established in a small town, probably easier than a large town with multiple practices. Practicing in a rural setting can facilitate knowledge of the community and provide a base of loyal patients. In a group practice, a young physician can learn from senior practitioners. This is something you would not have the opportunity to do in a solo practice. But this is not insurmountable.

Often, solo practitioners purchase a practice from a retiring doctor and can reorganize it to fit their needs and practice plan. Purchasing an existing practice allows access to an established patient base, and possibly trained staff, good equipment and infrastructure. This can facilitate a practice becoming rapidly operational and can be extremely cost effective.

It is important to have your lawyer review the purchase and conditions of the agreement. Equipment purchase can be negotiated. Your accountant may also need to review the account payable, receivables and financial health of the practice. An important purchase for a new solo practice is an electronic health record system. The best time to install the software is prior to opening your new office. This may avoid the costs associated with scanning charts and redesigning workflow if it is a new office. It does take time to become comfortable with any electronic patient software, but it will greatly help the efficiency of your practice and your access to your patient information.

Ultimately, as a solo practitioner, you can create the type of practice you envision, be in a location you like and provide the type of care you want. You can deliver more personalized care and not have to adhere to practice guidelines you may not be comfortable with, such as when in a larger group of doctors. You manage your own finances. In total, solo practices have many unique benefits.

Joining a Group Practice

Joining a group practice, either with a solo practitioner or with several colleagues, affords multiple advantages: seamless coverage of patient care, economies of scale by sharing expenses and resources, shared staff, space, equipment, supplies, phone, electricity, IT infrastructure, and so on.

In a group practice, there is the possibility for personalized patient care as everyone works as a team. Being with a team may bring additional benefits for your patients; for instance one colleague may have interest in sport medicine or in palliative care. The biggest benefit of joining a group practice is the ability to share information, to discuss complex patients, to increase the enjoyment of a busy practice with some camaraderie, to help with administrative decisions such as buying equipment, hiring new staff, to share call schedules and help provide seamless care to your patients during your vacations.

A happy partnership is a compromise and must include communication, discussion and planning. The first steps would be negotiating the conditions of joining the group, the use of space and staff, the process of billing, sharing expenses, the income, the on-call schedules, vacation time and patient coverage during vacation time or time away. These discussions should also include topics such as: the future purchase of equipment and expansion of the practice. To make no mistake, you should again do your due diligence by having your lawyer and accountant review the information before the final agreement. By clarifying all of this information, it will help you and your colleagues enjoy your workplace that much more.

Joining a Subspecialty Group

Being in a subspecialty group has a number of advantages that affect patient care. It allows all members of the group to fully develop their expertise in the specialized area of their specialty. A tertiary referral practice is likely to see larger number of patients with uncommon conditions allowing members of the group to gain experience that leads to better treatment outcomes. Such a practice will have staff and personnel who have developed specific knowledge and skills. Members of the group can call on colleagues to discuss complex cases. They can also share information gained at national and international meetings. Clinical results and clinical research data can be tabulated and presented and/or published.

Financial matters can be divisive because of the size of the group and the uneven seniority of different members with different income. Every measure should be taken to avoid conflict and keep the focus on patient care, education and research. Your lawyer and accountant should be very useful here as well in reviewing all aspects of the practice and contract agreement. It is not uncommon to have a year's probation at a decreased salary, benefits and responsibility to ensure that the new colleague (you) can deliver excellent quality of care, that you are responsible and professional, knowledgeable, and can blend into the practice well.

It is imperative to recognize the talent and value of each individual of the group, to allow certain specific patient care decisions be made on expertise and interest of certain members. Conversely, regular meetings with the group are needed to review issues in the practice, discuss decisions about purchases or hiring, to review mortality and morbidity, and ensure that data information on patient care is recorded appropriately for statistical analysis and quality assurance.

A subspecialty group practice allows its members to develop a high level of expertise in one area of their field, whereby they can become an expert and pursue their interests by being at the forefront of the science. This brings more expertise and potentially new treatment for patient care.

Joining a Multispecialty Group Practice

When deciding to join a multispecialty group practice, it is important to look at the composition of the group: internists, pediatricians or family physicians, maybe a surgical group practice such as a surgical clinic with a group of general surgeons, urologists, orthopedic surgeons and plastic surgeons. There are some of these types of private clinics in most of the provinces. They provide a combination of private care and government insured care.

Multispecialty groups can be set up in a variety of financial arrangements, such as ownership by the group themselves, ownership by an organization, or ownership by one physician who hires staff under contract. Interviewing for a group practice position, be it in a family practice or subspecialty practice, is an important process. You need to spend some time to assess the group, to meet everyone in the group in the practice. You should be prepared to ask questions and don't be shy about asking questions that matter to you. These questions could include: office space, workload, call schedule, equipment, conference and vacation time, office patient information systems, patient database for clinical research, income and range of income, benefits and maternity leave details. The more information you discuss openly, the less chance there is of a misunderstanding to occur. Because of the possibility of many variables in the practice set up, you will need help from your lawyer and accountant. Be very diligent about understanding the realm of your position, the incremental income and benefits, the short and long term arrangements of the daily practice such as space, staff, call schedule, vacation and research time, the possibility of buying into the practice at a future date and anything else you may feel will be useful in developing your career.

Joining a Hospital-Based Practice

A hospital-based practice can really be interpreted in different ways: a salary practice, a contract practice and a fee-for-services practice. A salary hospital-based practice is when you are hired by a hospital, mainly a university hospital, as a member of a division or department, for instance, as an oncologist, an intensivist, a general surgeon or a radiologist. Most university or university-affiliated hospitals across Canada will hire these specialists under a salary template with benefits up to 18%. These salaried contracts will include office space, secretarial staff, hospital patient information system, paid holiday and sick time, maternity leave and travel expenses. There may be issues negotiated on a personal basis.

A contract hospital-based practice refers to signing a contract type arrangement with the hospital. It can be a single person contract but usually these are groups of specialist contracts, such as a surgical group. These contracts allow the group to negotiate not only their salary, but also all other aspects of their practice (office space, workload deliverables, support, holiday time, equipment and others aspects that have been previously mentioned). The group will likely have to pay rent, their own expenses including secretarial support, office expenses and travel expenses. It is a negotiation process and again, even if you deal with hospital administration, chief of the department, it is important to have your lawyer to review the specifics of your agreement, this is a crucial step.

Choosing an Academic Career

The decision to pursue an academic career is shared by a minority of residents. The vast majority of our colleagues choose to pursue private practice, for a host of reasons. Both academic medicine and private clinical practice share tremendous rewards and fulfillment however. The vast majority of leaders in academic medicine and the majority of successful physicians in private practice have remained in these types of practices for their entire careers.

A major attribute of academic clinical practice is the opportunity to become extremely knowledgeable and skillful in a specific area of specialization. This opportunity not only permits academic physicians to embark on new innovative strategies for management of diseases but also ensures a continuous flow of challenging patients with novel clinical problems.

Academic physicians are offered tremendous opportunities to teach, including a variety of opportunities for direct observation, for lectures, for clinical supervision of students, residents and fellows, and for creating teaching and educational materials. These materials include articles, manuscripts, books, textbooks, seminars and conferences. Teaching carries the responsibility of providing a balanced learning environment for the students, residents and fellows.

All academic departments expect their faculty members to engage in some aspect of inquiry and investigation, although research is not always confined to full time faculty members. Translational research bridging clinical to basic research is becoming an important field for clinicians and faculty members. It is understood that academic physicians will be involved in service activities inside their department/ medical school, such as, committees, review boards national and international committees.

Academic centers will provide young physicians with startup funding, laboratory space, research mentors and collaboration with other scientists. The participation in basic science, translational and clinical studies, is essential for the advancement of medicine. In order to provide such an environment for young investigators, academic departments should provide protected time for research, subsidize the salary if needed, allow for time and support for the young investigators to establish and develop his/her research interest.

Duty hours and on call scheduling can be controlled with less on call responsibility to facilitate research work. Although most academic physicians have not chosen their career path based on financial remunerations, the income is getting more competitive with generous benefits and pension plan. A fringe benefit of the academic practice is the possibility after a set number of years, to have paid sabbatical time. This is nearly impossible in a private practice.

Academic practice lacks the autonomy of private practice. In general, decisions about office space, support staff equipment, are made by a network of professionals or by the head/chair of the department, within the university hospital. The academic physicians must be flexible and willing to work with a variety of people not always of his/her choosing. Nevertheless, academic physicians who develop a reputation for team building and respect can attract the best people in their practice.

Research has become increasingly costly. Competition for funding is fierce and federal funding for medical research is getting sparser. Physician-scientists must successfully compete against full time scientists for scarce research dollars and that is a challenge. Physician-scientists remain very valuable to the research community due to their understanding of translational medicine, the blending of clinical and basic research.

Academic physicians are expected to publish their work and report their activities at national and international meetings. Rarely is there adequate protected time for preparing grants, writing manuscripts and doing research. This is why it is necessary to have protected time in your contract. It will save you when the clinical practice and the research both get busy. Attendance and participation in national and international meetings and conferences are a fundamental part in building a reputation.

Achievement of "tenure" status is a critical goal for non-clinical university faculty members and there is substantial effort invested toward obtaining this appointment. It conveys a level of prestige and it is as a reward to research excellence.

A Medical Career in the Armed Forces

Although military practice is different in some respects than civilian practice, the fundamental role of providing compassionate and competent care to the patients remains the same, whether they are in uniform, family members of people in uniform, or have previously worn the uniform.

Choosing to be a military physician and spending a portion of your professional career in the military is not to be discounted. It may be a good opportunity especially if your education is being funded by the military. Military physicians serve as active duty officers. It is not uncommon for a physician to sign up for a few years and then decide to spend 20 years or more on active duty.

Military physicians are presented with unique training and practice opportunities. These include; training and experience in aviation medicine, dive medicine, undersea medicine, jump/parachute training and deployment medicine. Training for and practice in a simulated disaster and multiple casualty exercises, is also an integral component of the curriculum. The opportunity to travel and be exposed to a wide range of locations, culture and settings, are just some of the major benefits of military service. If you do consider work as a military physician, you should be willing to let a "travel agent" determine where, with whom, and for how long you will travel. Among the multiple opportunities for career development, are training and assumption of command responsibilities, managing people and organizations.

Compensation for physicians, particularly specialists, is not on par with what is currently possible in academic and/or private practice. Nonetheless, service in the military offers opportunity for lifetime retirement pension after 20 years of service. It is a great pension fund plan, as it is independent of the vagary of the stock market. You can then use your pension military money for investment, for establishing a private practice or for personal use.

However, you should be aware that there are several pitfalls. If you do graduate from a medical school in a fully funded Armed Forces Program, you become eligible to begin a residency but depending on the Armed Forces needs, you cannot assume that they will automatically be able to seek specialty training immediately after graduating. There may be a variable amount of time of active service before starting your specialty training. Although with the current training requirements and regulations of the Royal College of Physicians and Surgeons of Canada, you may need to complete 2 years of Family Practice Residency to obtain a MD license before contemplating specialty training.

This particular aspect of our current Canadian medical school curriculum program, whereby a trainee can continue on to a 5-6 years specialty program directly after graduating from 4 years of medical school, will need to be reviewed and assessed by the Canadian Armed Forced Health Services. Other pitfalls include the assessment of one's ability to tolerate unexpected and often undesired assignments. This is a major part in the decision-making process regarding a potential military career, over and above the funding support. You should ask yourself and your spouse or partner, if it would be comfortable to be relocated to remote locations or to live in active combat zones.

Failure to recognize this very real possibility can lead to a severe onset of emotional stress. Military life is highly regimented and hierarchical, which for some people can be a poor fit to their personality. Other disadvantages include job changes, family moves, and decreased physician's autonomy. Some individuals will experience personal distress. Attempts to seek release from previous commitments made with the Armed Forces may be difficult and have legal implications.

General Attitude

What you should have learned in residency can be summarized under 3 points: availability, ability and affability.

Availability is essential particularly at the beginning of your practice, wherever your practice may be. You need referrals, you need to build your practice, and you need to establish a name for yourself. Therefore extra time answering phone calls, talking to colleagues, accepting patients without appointment, spending extra time with particular patients are all building blocks. It may not be the approach you had in mind at the start, but these steps will help you and bring rewards. In fact you should keep this attitude even when you have a busy practice. The benefits and rewards will be many.

Although you know you have been trained well and you may feel confident in your skills, the people and staff around you need to get to know you, your personality, your abilities, your skills, and your attitude toward work. There is no need to be the "know it all." Take advice, listen, introduce yourself to as many people as you can and share your knowledge. Again the rewards are often multiplicative.

Affability seems easy to understand but it is very common that under stress, in difficult situations, medical or surgical, one can lose his/her control and be abrupt or impolite. This behavior ultimately accomplishes very little or next to nothing. Yes, we are human and not perfect. Under stress, we can be nervous and people/staff around us do understand that, but we should always make an effort to be respectful. Your reputation in the hospital is something that is assessed by all. It is important to develop good relationships with colleagues, staff, nurses, administrative managers and referring colleagues, always take the time to be polite.

REFERENCES

1. K.K. Shaw, J.K. Raj
   The Ultimate Guide to Finding the Right Job After Residency
   McGraw-Hill, Medical Publishing Division, NY, 2006

PART 3: Support Personnel and the Realities of Practice

CHAPTER 7: The Difficult Task of Hiring a Secretary

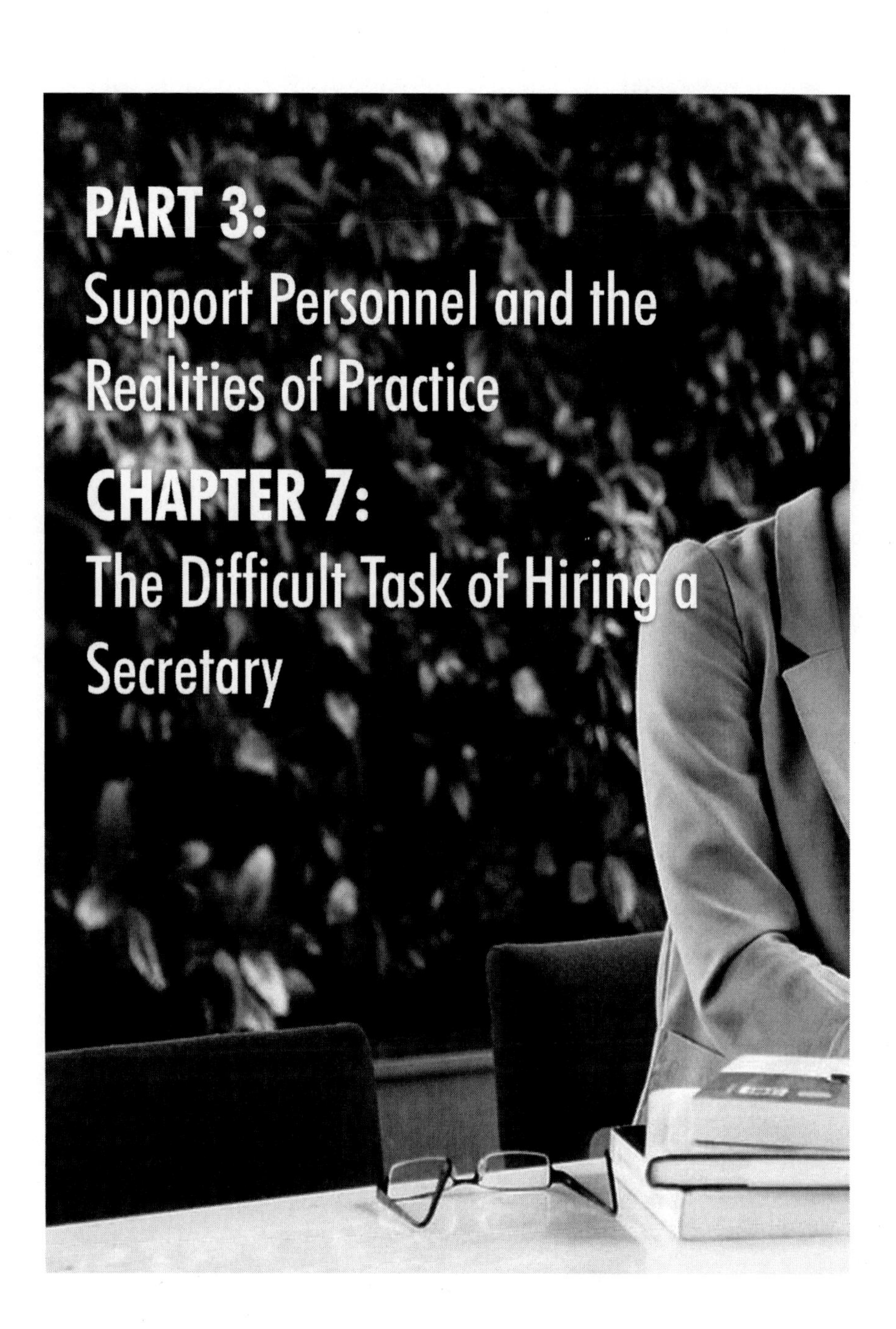

Planning to Hire

Hiring a secretary is a big step. The secretary is the frontline of your office for your patients, clients, sales representatives, visitors, etc. and insiders (hospital staff, colleagues, administrators, etc). Making a first impression is always an important aspect of establishing any relationship. How many times have you heard a colleague saying: "Wow, his secretary is very difficult to deal with" or a patient saying, "Dr. Smith's secretary is very abrupt." If you haven't, trust me when I say there is absolutely no need for this kind of attitude and it pays larger dividends than most realize to have a secretary that provides polite, affable, compassionate and helpful service. This should be at the top of the personal skill set.

Your secretary needs to represent you and your office in a professional manner and be knowledgeable about all facets of your practice, your services and your organization.

It is a demanding job with many requirements. You may get apprehensive when trying to hire the right person. No one wants to go through the process of hiring several people before finding the right fit. It adds a lot of stress and time that can be better spent taking care of patients. If you follow a few rules, you can hire an excellent secretary for your office with minimal difficulties.

There are many types of secretarial support arrangements you may encounter when establishing your practice or taking an academic position:

1. Secretarial/administrative support provided by the university:

   You may accept an academic position such as head of a university department. In your contract with the university, you should be provided with secretarial/administrative support. Many times, there is already a person in the position and she/he will be assigned to you. Sometimes, you may have the possibility to interview a new person but under university union agreement/salary/benefits. Although technically this person is your secretarial/administrative support, he/she is still an employee of the university/province.

2. Secretarial/administrative support provided by the hospital:

   You may accept a hospital based salaried position such as a: radiologist, an intensivist or an emergency physician. You should be provided with secretarial/administrative support. Again, you may be assigned a person that was already in the position or sometimes you may have the opportunity to interview others, but these people still fall under the union employee contract of the university/province.

3. Secretarial/administrative support provided by your practice group:

   You may join a group practice and they may have a team of staff that can help you start your practice. If not, you will need to hire your own private secretarial support.

4. Hiring your personal secretarial/administrative support:

   You are establishing a private practice, joining a group practice or opening up your own private practice. The following information will be helpful in hiring the best secretary.

Job Application Process

Look at the candidate's work experience to determine if he/she has held previous secretary positions. Refer to how long the candidate commits to jobs (if it is not a new person coming out of training). If you are looking for a long-term secretary, you may want to look at a person with limited experience that you can teach. This person may be able develop the skills to meet your needs. Of course, the training period will be shorter if you hire a person with experience, but this person may not be with you as long.

Contact the candidates that you are interested in and ask them to come in for a job interview. Provide a date and time for the interview. During the interview, don't just ask interview questions, check out the feeling you get from the individual. For example, she/he should smile, appear friendly and welcoming, be well groomed, and use proper grammar. Never underestimate the details.

Contact their references that are listed on the candidate's job application or resume. Ask the referees questions to determine the kind of employee the individual will be. Hire the secretary who makes a good interview impression and receives positive feedback from his/her references.

Secretary Skills

The skills of the person you hire is the most important aspect of your hiring decision. You may really click with the individual but if he/she cannot perform the important tasks required, it may not work for you. The exception here is if you really want to take the time to teach the person. When you create the skill set list, you should include every possible task he/she could perform, even if it is unlikely that they will ever use this skill.

Computer Skills

Secretaries must have solid computer skills. Those skills include the ability to use a variety of office software programs and adapt quickly to software that they are inexperienced with. It is important for the secretary to be open to learn new skills.

Communication Skills

Secretaries need to have solid communication skills, both written and verbal. Secretaries are often asked to communicate on behalf of you and they need to be able to present himself/herself and yourself, in the best possible light.

Secretaries are often asked to compose letters with minimal instruction. They must have excellent grammar and writing abilities. The need for excellent communication extends to all areas, from relatively informal communications like email and instant messaging, to research/administrative proposals and formal letters.

Equally important, your secretary should have a neat and clean appearance. That does not mean that they have to wear the latest fashions however. Another underrated yet important trait, is important is that they are cheerful and have a sense of humor. Office work is sometimes difficult and she/he will need to stay calm under stressful conditions to contribute to a positive attitude.

Typing and Administrative Skills

Secretaries need to have very strong typing skills, a good portion of their day is spent typing letters, memos, emails and other written communications. You may want to consider asking secretarial applicants to take a typing test.

Many secretaries also perform administrative functions, like basic accounting and payroll. It is very important that he/she is well organized, aware of details and has the ability to make other's jobs easier. It is no small task but it does help that your secretary is able to help the people around him/her.

People Skills

Strong interpersonal skills are essential for any secretary or administrative assistant. These individuals routinely work with patients, families and a number of different departments and they need to be able to get along with everyone in the organization in order to do their jobs effectively. When dealing with families, your secretary despite a busy schedule, needs to be able to listen to patient's issues, complaints, stay calm and composed, and be able to reassure patients. A friendly attitude and the ability to get along with others is essential for the job.

Interview

i) Computer Skills

Discuss the prospective secretary's experience working with computer software applications. For example, secretaries typically need experience with MAC, Microsoft Word, Excel, PowerPoint and electronic mail applications.

Should you require a secretary with advanced skills in other software applications, such as Photo Shop, solicit examples of jobs where he/her used these applications. You could ask the secretary in the interview to describe a time when he/she created a multi-page PowerPoint presentation with bullets, tables and notes. This will help you to discover the secretary's level of expertise in using computer software applications.

ii) Project Management Skills

Ask the prospective secretary to describe projects where he/she has played a supporting role. For example, you could ask a prospective secretary to describe how he/she prepared overhead slides or PowerPoint presentations for a quarterly departmental budget meeting. Inquire about steps that the secretary took to ensure the report was done on time for the presentation. Ask if the secretary made recommendations to senior managers that helped the project get back on track. These questions examine the secretary's leadership and independent thinking skills.

How to Select a Secretary

Establish the needs assessment for your practice. This will give you a good idea of what the secretary's job responsibilities should be. Ask other staff members to do the same so that you have a well-rounded picture of what administrative responsibilities are needed.

Look for standout skills on the candidates' resumes. You will need a secretary with solid writing skills, including proper grammar and spelling. Look for correct spelling and grammar in resumes and cover letters.

Hold interviews with candidates to get a good idea of how their experience, skills and overall personality fit with the needs of the position. Discuss their willingness to learn new skills and their ability to learn fast. Introduce them to other people in the office during the interview. This will help you gauge how well their personality will match to the people already working in the office.

Note his/her disposition in the interview and gauge how open and friendly he/she is while chatting. Also look at her appearance to make sure she understands how to physically present herself in a professional manner. When you check her references, inquire about punctuality, efficiency and any other standout skills.

Select the secretary that presents himself/herself as the most well rounded, experienced, and easy to work with. If you select a secretary based on skills alone, you may be miserable if he/she has a bad attitude. You may find yourself quickly looking for a replacement, resulting in a high level of stress.

Salary, Package and Benefits

In order for you to make a decision about the salary, package and benefits, you should ask other colleagues, to get an understanding of the average pay and benefits for secretarial support and what the positions entails. This will help you decide what you can offer in relation to your predicted income or current income.

You should write a contract detailing what the tasks will be, what you expect of him/her, what are the current and future plans of your practice or group practice, and what new skills may be required of him/her in the future.

You should offer a competitive salary, and an incremental salary plan over 5 to 10 years. Benefits can be more difficult to decide on and you should take advice from colleagues already in practice.

Benefit packages could include paid holiday time (2 weeks the first 4 years, 3 weeks the next 4 years and 4 weeks after 8 years), paid sick time up to 4 weeks, paid maternity leave up to 3 months, dental plan and perhaps a contribution to RRSP as a pension plan.

All of these benefits can, of course be negotiated with your new secretary and can be incremental over years. It is important to discuss the working hours and perhaps be flexible if he/she has young children to attend from time to time. There may be a need for overtime at some point, and you should address how it will be handled.

At the same time, you should discuss his/her long-term goals, such as work goals, personal goals, family plans and commitments. You do need to plan how your office/practice will work in the absence of your secretary (in the event that they need to take time off for holiday, sickness, maternity).

Employer's Attitude

As an employer and as the owner of your practice/group practice, you have a responsibility toward your employees, as you do toward your patients, colleagues and any other people you may interact with.

You need to always be respectful, professional and kind. You need to have clear communications about your requests, tasks, work you expect from your secretary, but you should maintain a trusting and respectful attitude towards him/her.

There is no need for conflict and despite stress in your practice, with difficult patients/cases or workload, you should remain considerate toward your secretary and the same applies to him/her. The day at the office will be much more interesting, enjoyable and ultimately a less stressful environment if you have a mutual respect with your secretary. A little appreciation will go a long way.

PART 3:
Support Personnel and the Realities of Practice

CHAPTER 8:
How Useful Is A Banker?

Better access to credit is one of the many benefits obtained by those with good banker relations. The biggest variable in any loan request is the person who is asking for the money.

As credit is more than just loans, a good banker relationship can also result in facilitating any requests you may have.

Establishing Relationship

Clear, frequent, open lines of communication are a necessary component of a strong doctor-banker relationship. Doctors and bankers should communicate at least twice a year. There is more involved in this relationship than mailing out your financial statements. Identical to any relationship based on trust, this one requires time. Invite your banker to meet with you and your financial advisor and review your 5-year plan. It will certainly make your banker understand your planning and your future needs. Create a good relationship and then capitalize on it later. The banker can be a friend, ally and consultant, but not someone in whom you necessarily confide, specifically about things that don't directly affect the banker's interests. If your marriage is on the rocks, don't rush to tell your banker. If something bad happens in your practice, wait and try to determine the cause and remediate if possible, before telling your banker. It is not hiding information, but it is unrealistic for a banker to want to know everything that is happening with you. It does not help your credit or transactions.

Finding a Bank

Pivotal to establishing a good banking relationship is finding the right banker. Most people will not have any trouble selecting a reputable bank. People prefer to use a bank that has a branch close to where they live or work, so they can bank with ease.

A national or international bank is a useful choice when you travel as you can have easy access to your money. You can transfer money to wherever you need to. Some banks do not charge any fees but, in general, there are some fees here and there inevitably, and you should inquire what the fees are and for what products/transactions. Deal with officers as high up in the organization as possible to establish a trustworthy relationship and have a banker who can address your needs without having to ask a higher-ranking banker for permission regarding your transactions. One stop shopping will speed things up and save you time. We will review the benefits of a dedicated banker or private banking services later in this chapter.

Different banks have diverse features, and even distinctive checking accounts. Know your needs and discuss them with a bank manager or a banker you have been referred to for their help.

For instance, most people will have a personal/family account and a savings account plus a business account and a savings account. Your personal/family account will be used by your spouse and yourself for depositing your income and for the daily household bills and needs. The personal savings account can be used to put aside monthly deposits toward the RRSP contribution and for emergency money. The business account will be used for practice income deposits and to pay for the practice expenses. The business savings account can be used for money you do not need immediately or as emergency money if you get sick or have an urgent need for extra cash.

The most important point to stress is the importance of having a long-term relationship of trust, mutual respect and communication with your banker. There will be several times in your life when a personal banker will be very helpful with house mortgages, office loans, car loans and even money management.

Simple Principles

There are 4 important points to understand and follow:

1. Build a trustworthy relationship with your banker and show the quality of your character. A relationship takes time to grow and only gets better if you contribute to it. With honest, frequent communication and time, a strong trust will develop between you and your banker.
2. Have cash flow and liquidity because when a bank gives you a loan, their main issue is the repayment and the collateral you have.
3. Your credit history, personal balance sheet and debt are important pieces of information for your banker.
4. Having real estate assets can be very useful as collateral for your loans.
5. If you have money, you have influence with creditors. If debt is kept to a minimum, it is possible to deal more easily with creditors/banks and not let them have the upper hand.
6. Banks want your deposits, your business. They know that as a doctor, if they loan you money, they will be repaid. Physicians are people of character and integrity. Doctors have a considerable income, cash flow, liquidity and also personal credit. Banks make money by charging fees for various services, loaning money and charging interest. You may be charmed into borrowing more money than needed but if you are careful, you may need far less. Despite the banker's charm, if you cannot repay your loan or are late in payments for any reasons, be assured that they will be talking to you to ensure that you pay. That would be the time when your relationship of trust and respect with your banker may be helpful in sorting out your problems.

Every word counts when signing a loan document: logic suggests that every word is there for a reason, even in the fine print. Do not allow any questions to be brushed away or with an answer that you are not satisfied with or not able to understand. Sign the note only when you have an understanding of the terms of the loan/note. Some of the terms that may be included in a loan document include requirements for collateral and the possibility that the bank can "call in" the loan. This refers to the bank having the option to demand an accelerated payment if they perceive you are in financial trouble, or for their own internal reasons.

A banker in general is not a financial advisor. A banker can only provide so much advice about investments. It will be ultimately be your responsibility to be aware of any terms and conditions, as well as the market.

A failure to communicate can cause a disconnect and can cause a misunderstanding between you and your banker. You may be unrealistic about your financial position and/or the banker may fail to explain his/her side. Banking is complex and highly regulated. Make sure you understand the banker's answers to your questions.

Determining the Products You Need

Over the years, various banking products have been developed as an outgrowth of the bank's role in financial intermediation. Many years ago, few types of banking products and services existed. Over time, the number and variety of products and services have increased. Be sure to understand what you need and not be offered/sold that which you do not need.

We use transaction/chequing accounts for daily expenses as the funds are easy to move in and out and cheques are still a widely accepted method of payment. There are fees attached to a daily regular chequing account.

Inquire about the cost as it may differ from bank to bank. Also, the interest paid on a transaction/chequing account is usually a low rate. Savings accounts are interest-earning and usually have few restrictions on deposits and withdrawals. Regular savings accounts usually pay a low rate of interest and require a minimum balance. It is common for a person to use their regular savings accounts for emergency funds and to supplement the funds maintained in a chequing account.

Certificate accounts are accounts that typically require a higher minimum balance and offer higher interest rates for a fixed period of time or term. Interest rates are often fixed for the term and therefore produce a predictable return. There is usually a monetary penalty on early withdrawal.

If you redeem the certificate before the end of its term there is usually a penalty.

Loans

Loans can be classified into several categories:

1. Short-term loans typically have a term of less than a year and may be used for your practice purpose or personal use.
2. Long-term loans are for periods longer than a year and may be used for various purposes including expanding your practice, buying equipment or buying personal items such as a car. They are usually repaid in monthly or biweekly installments.
3. A line of credit is essentially a pre-approved credit limit against which you can borrow. You can repay the borrowed fund in installments within a certain limit of time or you can repay in a lump sum.
4. Consumer loans can be repaid as installment credit at a fixed interest rate and a fixed interval repayment. A mortgage loan is for buying a house and there are multiple types of mortgage loans. You will need to do your homework to find the one that is appropriate for you.
5. A homeowner can also obtain a home equity loan (a second mortgage), against the portion of your home's value and use the money for education, car or renovations.

There may be one or more times in your life when you will be asked to help a friend, a colleague or a family member by co-signing a note. Your desire to help may cause you problems in the future. If your friend, colleague or family member cannot pay the money back or meet the payments, the bank will go after the person with the money and that could be you, because you co-signed the note. You are fully liable and responsible as the co-signer so go in with your eyes open.

"I had a personal friend who was working very hard in an upscale food store and catering business. She wanted to buy the catering part of the business and developed a line of specialty foods.

She wrote a business plan that sounded solid, with all the appropriate information. I helped her by co-signing a business loan. For years, there were no problems and the monthly payments were made on time until my banker called me to report that my friend had not paid her monthly payments for the past 3 months.

I reached my friend who assured me she would pay. Only one payment was made and none thereafter. She avoided my phone calls for several weeks and finally declared bankruptcy. I became liable for paying back the loan, as there was no point in suing her."

You may want to help your best friend, colleague or family member, but you must exercise caution and be prepared for the worst. That, or simply don't take on the risk.

Insist on the lowest mortgage rate, or the lowest interest rate possible for your loan. It can always be negotiated, as there are many other agencies that will be happy to get your business. Always verify mortgage rates and loan interest rates before starting to negotiate with your banker. It will come in handy and provide you with the best option to get the best rate.

On-line banking is now common and allows you to use your computer to access your accounts through a secure password protected on-line system. Make sure to use a safe banking Internet protected system to guard your account information. It is useful to know how to do on-line banking for paying credit cards, transferring money and so on, but it does not replace a personal banker for important transactions such as mortgages and loans.

Student Chequing and Savings Accounts

For a student, there are some considerations before choosing where to open an account. There are many excellent banks in Canada and they all offer slightly different products that you should familiarize yourself with. Some offer a number of free transactions per month, free cheques, air miles and/or line of credit. I recommend choosing an account with a no-fee chequing account and a savings account of 1-2% interest.

Student Credit Card

Most people think they cannot live without a credit card, as it appears it is a 'must have' in our society. It is useful to have a credit card but you need to control your spending. Many students will want to consider applying for a credit card with a limit of at least $500. Unlike personal bank accounts, this is where your credibility for future mortgages or bank loan is built. You also have to think about that kind of rewards you want to get out of your card: travel, cash back and points.

Standard interest rates for credit cards hover at about 19-20% per year, so avoid holding debt for more than 20-30 days and pay the balance in full every month. Never max out your card, in case of an emergency. I recommend choosing a card with no annual fees and the maximum amount of perks. If possible, do not have more than one card until you start your practice.

Student Lines of Credit

You may need some extra cash and you do not qualify for a provincial student loan. You might want to consider a student line of credit. Banks usually make these loans at around the prime interest rate. You will need a co-signer to vouch for your line of credit and your family may be able to help you with this.

Repaying your line of credit over time after graduation also helps build a good credit for future loans. It is recommended that you use this option for tuitions fees and other schooling needs with a low interest rate, rather than using your credit card. Again, different banks offer different products, rate and conditions. Do your due diligence in reviewing several of them in relation to your needs.

In summary, finding a banker and establishing a professional relationship will help your banking transactions, your needs for loans and mortgages. Your credit rating can overall enhance the ease of running your office. All these services come for a small cost in the context of your busy schedule.

PART 3:
Support Personnel and the Realities of Practice

CHAPTER 9:
An Important Person - Your Accountant

Accounting

Accounting is the measurement, processing and communication and filing of financial information to the appropriate agency. Accounting measures economic activities (for instance, your practice and your personal income) and conveys this information to a variety of people, the government tax office, perhaps your investor/financial advisor and banker. The terms 'accounting' and 'financial reporting' are often used as synonyms.

Accounting can be divided into several fields including financial accounting, management accounting, auditing, and tax accounting. Financial accounting focuses on the reporting of financial information, including the preparation of financial statements for people such as the government tax office, your financial advisor and banker.

Management accounting focuses on the measurement, analysis and reporting of information for internal use by management. The recording of financial transactions so that summaries of the financials can be presented in financial reports is known as bookkeeping, of which double-entry bookkeeping is the most common system.

The accounting of a single practice or a group medical practice is not very complicated in relation to other types of organizations. The bookkeeping for your practice can easily be maintained by your secretary and provided to your accountant at year-end.

Choosing an Accountant

- Look for compatibility – do your research into the firm, their values and their people. The important factor is to look for a practice or an individual that you can relate to.
- Be clear about the services and support that you require.
- Ask the firm if they have other clients that are similar to you - you're well within your rights to ask for references.
- Get recommendations from colleagues, friends and family.

- Ask for details around the fee structure to ensure that there are no surprises.
- For the best service, always choose a Chartered Professional Accountant (CPA) for the best information.

What should you expect from your accountant?

If you have been clear with what you require and your accountant has been clear with the services they can provide, then you should expect timely, relevant and insightful advice. Your accountant should:

i) Be straightforward, honest and sincere in their work
ii) Be impartial, intellectually honest and free from conflicts of interest
iii) Have a high standard of competence and complete all professional obligations with due care and in a timely manner
iv) Carry out all work in accordance with the relevant technical and professional standards of Canada.

Accounting Services

Accountant Duties Include:

- To prepare, examine, and analyze accounting records, financial statements, and other financial reports to assess accuracy, completeness, and conformance to reporting and procedural standards
- Prepare tax returns, ensuring compliance with reporting, and other tax requirements
- Analyze business operations, trends, costs, revenues, financial commitments, and obligations to project future revenues and expenses (budgeting) and to provide guidance to maximize profit and guide practice expansion through growth or acquisitions
- Develop, implement, modify, and document record keeping.

- Advise yourself and/or your group practice about issues such as resource utilization, tax strategies, and the assumptions underlying budget forecasts
- Provide auditing services as needed
- Advise yourself and/or your group practice in areas such as compensation planning, employee benefit design and pension/retirement planning
- Investigate bankruptcies and other complex financial transactions and prepare reports summarizing the findings (hopefully never required)
- Represent you before tax authorities such as the Canada Revenue Agency (CRA) and provide support during litigation involving financial issues
- Appraise, evaluate, and inventory real property and equipment, recording information such as the property's description, value, and location. He or she may enlist the assistance of a qualified appraiser
- Maintain and examine the records of government tax revenue authorities
- Provide financial planning and estate planning advice and serve as executor of an estate, if appropriate.

Records:

Well-kept records can help you plan for decisions that will affect the future of your practice. The more organized your records are, the more your records can help you in the future to make informed decisions.

Organized records can help you:

- Track and compare past and present finances, revenue and expenses
- Plan and forecast future financial positions
- Provide information to make sound business decisions

- Satisfy reporting obligations, notably to the Canada Revenue Agency
- Save time and energy if your practice gets audited.

The Types of Records to Keep:

When you run a practice/business, you need to keep records of various business transactions. These records include:

- Details of all transactions – usually contained in an accounting software package
- Paper or electronic receipts for all expenses and asset purchases
- Detailed record of all sources of revenue
- Details of expenses including payroll details
- Taxes collected and paid.

Records Retention

As a general rule, you need to keep all records and supporting documents used to determine your tax obligations and entitlements for a period of six years from the end of the last tax year to which they relate. At a minimum, the financial records should be permanent, accurate with a complete record of your daily income and expenses.

Styles of Bookkeeping

A useful record keeping system is one that is simple to use, easy to understand, reliable, accurate, consistent and accessible to provide you with information on a timely basis.

You can learn more about bookkeeping with the help of online tools, books, business services and software. Introductory training in accounting can also be helpful if you are unfamiliar with accounting processes; many local business service centers offer such basic training, as do colleges.

New doctors should use an accountant from the start of their practice. Your accountant and your secretary will become an integral part of your future decision-making.

Fees: Fixed or Hourly, and How Much?

A Chartered Professional Accountant (CPA) will provide an estimate of fees at the beginning of an engagement, most likely by the hour. A CPA will also provide a letter of engagement setting out the respective responsibilities of the client and the CPA.

Hiring an accountant at the beginning of your practice will save you time and money. Establishing a good relationship with your accountant will allow you ready access to valuable advice related to compensation planning, retirement planning, education funding for your children, tax minimization techniques and wealth creation strategies. In addition, your accountant can provide invaluable advice related to estate planning, providing your loved ones with peace of mind by maximizing the after-tax value of your estate and distributing it in an orderly and equitable manner and estate settlement with a plan for relieving your beneficiaries of the complex, time-consuming and stressful job of being the estate's executor/s and trust planning with creating long-term financial security by ensuring that the needs of your dependents are always met, in sickness or in health.

Income Tax

Reasons to file a tax return:

In general, if you owe tax, you are required to file an income tax return. If you do not owe tax, you still might want to file a return in these situations:

1. You owe tax or want to receive a tax refund to recover any tax you may have overpaid as tax installments or from taxes withheld by an employer or through your investments.
2. Take advantage of refundable tax credits. Examples of refundable tax credits: Child Tax Benefit and GST/HST Credit and provincial refundable credits. You will not receive these credits if you do not file a return.
3. Create contribution room in an RRSP by reporting earned income received during the year.
4. Recognize and carry forward or transfer any unused tuition, education or textbook amounts. If you don't use your contribution room in any year, you can carry the unused amounts to future years or transfer the credits to certain relatives.
5. Recognize any business or capital losses, which may be carried forward to future years or carried back to prior years.

Penalties for Late Returns

If you owe tax and you file your tax return after the deadline, you'll pay a late filing penalty – plus interest on the tax owing and on the penalty.

The first time you file late you can expect to pay:

- A late-filing penalty – 5% of the amount of tax you owe, plus 1% for every month that your return is late, for up to 12 months. That adds up to a maximum of 17% of the tax you owe.
- Interest – at the prescribed interest rate on the amount you owe, beginning on the day after the tax-filing deadline (generally April 30th). You will also be charged interest on any late-filing penalties.

CRA Tax Review

The Canada Revenue Agency (CRA) may contact you about your tax return, and not necessarily just at tax time. This is called a tax review — it is not the same as a tax audit. But it may lead to an audit if the CRA is not satisfied with your response to the review. This is one situation where a good accountant is important because he/she can easily handle the CRA request for information, particularly if they are familiar with you and your tax records.

Tax Audits

A tax audit occurs when the CRA examines your books and records to assess if you have paid all the taxes you owe. Generally, the CRA can audit tax returns within 3 years of the date of the original notice of assessment. If you're selected for an audit, it is in your best interest to:

- Co-operate with the auditor
- Answer all questions respectfully, and
- Provide any information and supporting documents requested in a timely manner.

Again, it is better to let your accountant handle this issue with the CRA and cooperate. The CRA has the right to audit anybody.

Registered Retirement Savings Plan (RRSP)

A RRSP is a registered retirement savings plan to which you contribute a certain amount per year as determined by your earned income. Deductible RRSP contributions can be used to reduce your tax.

Any income you deposit to your RRSP is usually exempt from tax as long as the funds remain in the plan; after you retire, you generally have to pay tax when you receive payments from the plan.

Setting up an RRSP:

You can set up a registered retirement savings plan through a financial institution such as a bank, credit union, trust or insurance company. Your financial institution will advise you on the various types of RRSP and the investments they can contain.

You may want to set up a spousal RRSP. This type of plan can help ensure that retirement income is evenly split between both of you. The benefit is greatest if a higher-income spouse contributes to an RRSP for a lower-income spouse. The contributor receives the short-term benefit of the tax deduction for the contributions. Your spouse, who is likely to be in a lower tax bracket during retirement, receives the income and reports it on his or her income tax and benefits.

You may want to set up a self-directed RRSP if you prefer to build and manage your own investment portfolio by buying and selling a variety of different types of investments; however, this approach can be time-consuming and most likely not your area of expertise. If you are considering RRSP's, be sure to consult with your financial institution or your financial advisor.

You make your RRSP contributions directly to the RRSP issuer, most likely your financial advisor or financial advisor institution. Your accountant can provide you with your RRSP contribution limit for that year.

Contributing to an RRSP:

The deadline for contributing to an RRSP is within 60 days after December 31 of every year. The age limit for contributing to an RRSP is December 31 of the year you turn 71 years of age. This is the last day you can make a contribution to your RRSP.

Withdrawing from your own RRSPs:

You can withdraw amounts from your RRSP before it starts to pay you a retirement income. However, you will have to pay tax on the withdrawn amount.

Receiving income from an RRSP:

If you are near retirement, you may be thinking about how you will receive regular income from your RRSP. You generally have a certain amount of flexibility on the types of income you can receive.

At any age up to the end of the year you turn 71, you can choose several options for your RRSPs, the most common being to transfer your RRSP funds to a registered retirement income fund (RRIF) or buying an annuity. Consult your accountant and your financial advisor to determine the best option for you and your family.

Tax-Free Savings Accounts (TFSA)

Tax-free savings accounts (TFSAs) are designed to help Canadians save more.

- TFSAs are available to Canadians age 18 and older.
- You can contribute up to $5,500 each year (2014 TFSA limit). If you don't contribute the full amount each year, you can carry forward the unused amounts, based on the contribution limits for each year.
- You can save tax free for any goal you want (car, home, vacation).
- You do not need earned income to contribute.
- You do not have to set up a TFSA or file a tax return to earn contribution room.
- You can take money out when you want, for any reason, without paying any tax.

- If you take money out, you can re-contribute it the following year, in addition to the annual $5,500 maximum. You can invest your TFSA money in a wide range of investments, like cash, GICs, bonds, stocks and mutual funds.
- You can give your spouse money to contribute to his or her own account.

Registered Education Savings Plan (RESP)

How RESPs work:

- A RESP is a dedicated savings plan to help you save for a child's education after high school.
- Most RESPs are opened for children, but you can open an RESP for yourself or another adult.
- When your child is in post-secondary education, they can start taking payments, called Educational Assistance Payments (EAPs) from their RESP. EAPs are made up of the investment earnings and government grant money in the RESP. The person who is named to receive EAPs under the plan is called the beneficiary.

RESPs are useful because:

- Your savings grow tax-free. There is no tax on the investment earnings, as long as they stay in the plan.
- If you save for a child age 17 and under, the federal government also puts money into the RESP as a grant or bond. In some provinces, the provincial government may contribute too.
- You can usually put money in whenever you want, up to a lifetime maximum of $50,000 per child, however some plans require set monthly or annual contributions.
- The contributions are not tax deductible, although you can withdraw them tax free from the plan at any time for any reason.

- There is a wide range of investment options available for RESPs, such as stocks, bonds, mutual funds and GICs.
- Your child can take money out of the RESP when they enroll in university or college or another qualifying education program or specified education program.

Companies that offer RESPs are financial institutions such as banks, credit unions, mutual fund companies, investment firms and trust companies. They offer individual (one beneficiary) and family (several beneficiaries) plans.

Tax Planning and the Family Trust

Under current income tax legislation, it is difficult to reduce income tax liabilities within a family group via techniques commonly referred to as "income splitting". Income splitting attempts to ensure that income earned within a family is evenly distributed for income tax purposes so that by ensuring that tax benefits allowable for each family member are utilized (i.e., low tax brackets, personal tax credits and capital gains deduction), the family's overall tax burden can be reduced. One of the few methods of income splitting still available is through a family trust arrangement. Although the effectiveness of income splitting via a family trust was significantly reduced by the introduction of the so called "Kiddie Tax" in the 1998 Federal Budget, important tax savings may still be generated via a family trust. Details are discussed below.

What is a Family Trust?

A family trust, in this context, is an arrangement whereby an individual may allow family members to share in the growth and value of an incorporated active business without that individual losing control over the operations of the business. In the typical situation, an individual (the "trustee" of the family trust), will hold property (shares of the active corporation) "in trust" for the benefit of family members (the beneficiaries of the trust). Although title to the property is in the trustee's name and the property (the active corporation's shares) is under the trustee's control, the income and capital growth attributable to the shares accrues to the beneficiaries. To the extent that income earned by the trust is paid to its beneficiaries, the income is taxed in the beneficiaries' hands. Where these beneficiaries earn little or no other income and are not subject to the "Kiddie Tax", they may pay little or no income tax on those distributions. Note that a typical family trust agreement is drafted so that the trustee may pay that income out to, or for the benefit of, any or all beneficiaries at his discretion as he sees fit. In that way, future problems associated with issuing shares directly to children can be avoided.

A family trust is established when a person referred to as the settlor, usually a relative, gives a gift to the trustee for the benefit of other family members. At the same time, a written agreement is drafted which sets out the terms whereby the trustee will hold and manage the property on behalf of the beneficiaries. This agreement gives the trustee the power to distribute funds from the trust at his/her discretion. The agreement is a critical part of any family trust arrangement.

A family trust can be used for income splitting and capital gains splitting. Under the most recent guidelines released by Canada Revenue Agency, the trust can pay for, or reimburse, a wide variety of expenses for a child as long as the payment of the expense clearly benefits the child. In all cases, receipts should be retained that document the fact that trust funds were spent on the beneficiary's behalf.

In summary, this is only a brief overview of the benefits and mechanics of utilizing a family trust in income tax planning. The use of competent professional advisors in establishing and maintaining a family trust is mandatory. This will help you avoid the many pitfalls associated with this complex form of tax planning.

Issues to Familiarize with:

- Different types of practices.
- Alternatives to the fee-for-service billing model.
- Pros and cons of incorporation.
- Planning for taxes.
- Finding employment, practice and locum opportunities.

The Canadian Medical Association's (CMA's) Practice Management Program (PMP) provides a variety of resources to help physicians make the most of their working life, from assisting you with setting up primary care networks to understanding the business of operating a practice. PMP also offers workshops on starting new practices, from a business perspective. MD Physician Services can help residents save time, optimize revenues and enhance patient care with billing, scheduling and Electronic Medical Record (EMR) software, consulting services, seminars, insured billing services and web portals for physicians and patients.

Practice types and remuneration:

As a physician, you have choices about how you would like to practice and how you would like to be paid. The following information is a brief overview of some of these options that are available to you.

Solo practice:

While a solo practice affords complete autonomy and dedicated resources, it also leaves you with the burden of all start-up costs and overhead expenses.

Group practice: Sharing office space with another medical professional of the same or another discipline can be a satisfying and cost-effective way to practice.

There are two primary types of group practice:

- Partnership: Share expenses, income and liability.
- Association: Share some or all of the practice expenses (as defined by an association agreement) without sharing income or liability.

Whatever you choose, protect your interests – have your financial and legal representatives review all contracts before you sign.

As a resident, you are paid according to the contract negotiated for you by the Canadian Association of Interns and Residents (CAIR). You may opt to continue the security of an agreed-upon salary (unaffected by the number of services performed) by accepting a position as an employee or contractor with a hospital or other organization.

Unlike a self-employed physician, you may not be able to deduct expenses from your income such as professional association fees or malpractice insurance or other expenses.

Alternate Plans:

Rather than billing the government health care insurance plan on a fee-for-service basis, you may opt for an Alternate Relationship Plan (ARP). ARPs provide physicians with set remuneration amounts in exchange for delivering:

- Services to a specific patient population.
- Services in a specific location.
- Services for a specific block of time.

Incorporation:

There are advantages and disadvantages to incorporation, which are discussed in more detail later in the chapter. Basically, you should consider incorporating your practice if you are currently earning income that is in excess of the funds necessary to support your lifestyle, you are currently maximizing your personal RRSP contributions, and there is enough time after incorporation and before retirement to make the deferral worthwhile. It will be necessary to estimate your annual personal lifestyle expenditures including annual RRSP contributions, annual income from your professional practice, and also to estimate/determine your desired retirement date.

Taxes:

Your tax situation may depend on whether you are self-employed or a salaried employee and the province in which you will be earning your income. Generally:

- Salaried employees will have taxes automatically deducted from their pay cheques by their employers.
- Self-employed physicians will need to calculate the taxes they owe, but may also deduct business-related expenses incurred to earn their income (e.g., continuing medical education, office expenses and professional association membership fees).

Other income may also be taxable, such as government benefits or investments. Consult your financial/tax specialist regarding your particular situation.

Medical Practice Incorporation:

Most provinces and professional associations in Canada now permit professionals such as doctors, dentists, lawyers, and accountants to carry on their professional practice through corporations. It is a subject that we, in going to practice, know little about and it may have substantial advantages for your practice, for yourself and your family.

When you are thinking about incorporating your practice, several questions arise:

- Should I incorporate my medical practice now?

- How do I incorporate?
- How does a medical corporation help me?

I cannot address all aspects of incorporation as it is complicated, however I will provide an introduction to incorporation.

What is Incorporation?

Incorporation is the creation of a legal entity called a corporation, also known as a company. A company is a separate legal entity. It can earn income, acquire assets, enter into contracts, pay taxes, etc. When a physician incorporates, the medical corporation will carry on the business of the doctor's medical practice.

Why Incorporate a Medical Practice?

There are many reasons for a physician to incorporate.

1. Income Tax Deferral

Deferral of tax is achieved by having the corporation retain a portion of the professional income where it is then subject to tax at a much lower rate than if earned personally.

For instance, in British Columbia, personal tax on regular income between $136,000 and $150,000 is taxed at 43.7% in 2014, and income over $150,000 is taxed at 45.8%. These numbers vary somewhat between provinces. Alternatively, in a private corporation, the first $500,000 of business income is only taxed at 13.5%, and the remainder at 26%.

A doctor can defer tax of up to 32% on income in excess of $150,000 and up to $500,000 by using a medical corporation if the funds are left in the company. This is an important point. Incorporating your medical practice will provide you with the best savings if you do not need to use all the company income for your lifestyle. For instance, if your practice earns $400,000 and you need all the after-tax money for your lifestyle, there is no deferral advantage in being incorporated. However, if your practice earns $400,000 and you give yourself a salary of $200,000, then your personal salary of $200,000 is taxable at personal rates as noted above, and the remaining corporation income of $200,000 is taxable at 13.5%. It is a significant tax advantage. If you need more money at the end of the year, you can ask your accountant in cooperation with your financial advisor to issue a dividend to the amount you need. Dividends are taxed at a lower tax rate and there may be an advantage.

2. Income Splitting

Canada uses a graduated income tax rate system where income at different levels is taxed at different rates. Income splitting is shifting income from a high tax rate individual to lower tax rate individuals. For example, two people earning regular income of $100,000 each will pay approximately $18,000 less tax combined than one person earning $200,000.

Unincorporated doctors can split income with family members by paying them a salary. However, an expense must be reasonable to be deducted from business income. A salary of $50,000 to a spouse who has little involvement with the medical practice would probably not be considered reasonable, however, income splitting to a maximum of $30,000 with your spouse, may allow you to decrease your personal salary from your corporation, provide the same total household income while decreasing your personal income tax and increasing the money left in the company.

The advantage that a medical corporation has over an unincorporated physician is that it can also pay dividends to the shareholding physician and family members. Dividends are paid from the after-tax income of a company. Dividend income is taxed personally at a lower rate than regular income and takes into account that some tax has already been paid by the corporation.

The Income Tax Act contains many rules related to income splitting, some of them designed to allow income splitting and many more to limit income splitting. You should consult your accountant on these matters.

3. Other Considerations:

Incorporating a medical practice can also have a positive effect on spending habits. When not incorporated, medical practice income is deposited into personal bank accounts which is easy to access as a single account for all business income and expenses as well as personal expenses.

A corporate-owned permanent life insurance policy can help you mitigate income tax obligations and maximize the wealth-building potential of your assets because investment earnings within the policy are tax sheltered. More on this subject is explained in the insurance chapter.

There is a process to incorporating your medical practice and with the right knowledge, it will not be overwhelming.

This includes the following steps:

- Choose a corporation name and make a name reservation with the provincial corporate registry.
- Determine shareholders, share structure and overall structure in the event a family trust is also used (handled by your accountant).
- Application to the College of Physicians and Surgeons for approval to incorporate.
- Prepare an incorporation agreement and articles of incorporation (handled by your lawyer).
- Apply to the College of Physicians and Surgeons for approval to incorporate.
- File an incorporation application with the provincial corporate registry.
- Submit incorporation documents to the college for your medical permit.
- Open a corporate bank account and pay for shares.

After incorporation, other administrative start-up tasks include registration with tax authorities, notification to the medical provincial authority, and other revenue sources that the practice is incorporated and notification to insurance providers, etc. Most of these steps that appear time-consuming can be organized through the assistance of your lawyer and accountant.

A professional corporation may provide significant tax benefits for you if your professional practice generates enough income to make the tax deferral worthwhile. Many other facts and circumstances will need to be considered by you and your professional financial, tax and legal advisors to determine whether a professional corporation is right for you. Your advisors will also be able to determine if there are any other benefits of incorporation available for you.

In summary, there are multiple ways that a good accountant can help you organize your practice accounting, your personal accounting, assist with practice incorporation, provide information and guidance on saving taxes, establish tax-shelters, establishing education savings funds and a family trust, prepare your financial portfolio for retirement, determining the best opportunities for using your investment income during retirement and many more tax-related issues that are beyond the scope of this chapter.

But one very important point is to choose your accountant carefully, with good references and clearly explain your needs and goals. Build a relationship based on trust, respect and professionalism. This will be to your benefit.

REFERENCES

1. www.cma.ca
2. https://mdm.ca/index.asp

PART 3:
Support Personnel and the Realities of Practice

CHAPTER 10:
Searching for a Financial Advisor

Physicians are highly trained professionals but receive little or no instruction on how to manage their money, invest and plan for their financial security.

The information we get about investing usually comes from our parents (some parents, of course, do not provide instruction), friends, colleagues, and from our own readings. The average medical student graduates with tens of thousands of dollars of debt. Add other potential expenses such as getting married, buying a new car, a condo or a house, and establishing your practice. By the time you are in your mid-thirties many physicians find themselves with a substantial amount of debt.

As a physician starts their practice, income can triple or quadruple in an instant; yet their money management skills may not match their increased income.

Doctors have extensive training and years of accumulated knowledge, save lives and treat sick people. They have high self-confidence and despite little knowledge about investing, they often think that because of their soaring income they will be fine.

Why?

In their career, physicians will only be able to make a limited amount of income. There are only so many hours in the day and so many patients a doctor can see. Because of this, there is a limited amount of income that physicians can earn. Sound investment throughout will help increase a physician's wealth and financial security.

Physicians make decisions every day and results are measured in hours, minutes or even seconds. A financial advisor can make decisions over days or sometimes months at a time. Do not rush over important investment strategies.

Who?

Finding a financial advisor is no easy task. Allowing a professional to manage a portion of or your entire portfolio can be extremely profitable. Using their knowledge and expertise, your financial advisor will help you to make specific investment decisions.

Many physicians would be well served by working with a wealth management or financial advisor at a brokerage firm, a large bank, an investment bank or a large private investment firm. The usual annual fee is around 1-1.5% but usually can be negotiated in relation to your portfolio and your needs. Not only will you be working with legitimate people, but also you will have the best chance to maximize your returns. More importantly, such advisors should help you avoid investments that are altogether too risky or inappropriate.

You should not lose control of your money. It is not a good approach. You should remain involved in every decision regarding your investment. There should not be any investments made without your approval and you should meet your financial advisor twice a year to review your investments, assess problems and plan for the next year.

So then, how do you choose a financial advisor that you can trust, establish a long-term relationship with, and discuss your needs and your long-term plans while remaining involved in the decision-making?

The best way to find professional investment advisors is to use the same decision-making process you use when evaluating a patient. Take note of history, examine the patient/firm, order tests/reading and obtain consultation/colleagues' advice. It is recommended that you consider a financial advisor working with/for a reputable institution. Canada has a large number of financial banks and investment institutions.

Useful Tips that Can Help

1. Assess your financial situation, find out what you own, what debts you have and what income you have. Determine how much cash you have available, your bank accounts, and your investments (if you already have investments).
2. You should set your investment goals: for example, buying a house (if not already done), mortgage, paying for your children's education, deciding on different types of investment, how much money to invest each year and planning for retirement.
3. Research financial advisors in your area. Talk to friends, colleagues, family and financial firms you already know and trust, such as your own bank or accountant. You can also search the area/town on the Internet.
4. Then set your eyes on a few people, contact them and discuss your situation with them, outlining your goals and your philosophy.

You have to understand that different financial advisors have different areas of expertise, so you need to determine the area of expertise you need for your future plans: conservative investments versus more risky investments, for instance.

Make sure to understand the qualifications of the financial advisor you are considering. Once you have found a financial advisor who appears to meet your requirements, set up a meeting of introduction. Any reputable financial advisor will hold an introductory meeting at no cost and no obligation for you. Bring all pertinent information to explain your needs and ask questions so you can determine if the financial advisor is the right person for you. You will need to ask what is his/her commission is. Is it, an annual fee (anywhere from 1 to 2% typically) or purely commission based?

The downside of commission is that you pay for any changes in your investment portfolio. An annual fee schedule is more popular these days and allows you more flexibility in your portfolio.

When you understand who your financial advisor is and how they will be compensated, then you can negotiate the rate that is suitable for you. I believe that the most advantageous scenario as your portfolio gets more complicated and diverse is to negotiate an annual flat fee for all your needs.

You need to be comfortable with the financial advisor you choose as he/she will need to know a lot about you, your family and your finances. You have to be comfortable enough and trust them enough to share substantial personal information. You do not want to put your finances and hard-earned money in jeopardy.

Building Trust

In your discussions, the financial advisor should be able to answer your questions and demonstrate his/her advice is based on what is "evidence-based investing" information about markets, stocks, mutual funds, bonds and commodities with a historical review, not just a personal opinion.

This is a bilateral relationship. It will allow you to learn more about the markets and understand the movements (both up and down) of your portfolio investments.

One more important point: ensure a seamless communication between your banker, your accountant and your financial advisor. All of them deal with your money and should ultimately maximize your investments, be it the best mortgage and loans rate for you, the lowest income tax payments, the greatest investment returns and sound retirement planning.

You may hear that a financial advisor may be "independent". This may be true or not completely true. Investment firms are regulated and allow brokers/financial advisors to call themselves independent if they work as independent contractors with an investment firm.

However, these type of financial advisors are still limited to sales activities permitted by whichever firm he/she is affiliated with. They are still held to the same standards of their firms and their work contract arrangement. You should know and be comfortable with their standards and know that they will not interfere with your relationship and investment planning.

In the chartered banks of Canada, there are banks, credit unions, and trusts. Banks include domestic banks, Canadian banks that are subsidies of foreign banks and foreign banks with branches in Canada. There are government-owned banks, such as Bank of Canada, Business Development Bank of Canada and Farm Credit of Canada. Canada has a strong Credit Unions financial services sector, such as Vancity, Coast Capital Savings, and Caisse Populaire Desjardins in Quebec. There is also a large list of investment institutions in Canada that you can access and review online.

MD Management of Canada

MD Physicians Services is a wealth and practice management firm geared to meet the needs of physicians. (*mdm.ca*)

Owned by the Canadian Medical Association, it helps physician clients to build wealth and financial security. With close to $30 billion in assets under its management, the group has solid expertise and experience.

MD Physician Services is the manager of the MD family of funds, a group of 13 mutual funds. The company retains and supervises investment managers, who manage the portfolio assets of some or part of the mutual funds. It focuses exclusively on helping CMA members. MD Management offers full financial planning services through its many branches across Canada, allowing investors the possibility to delegate the day-to-day investment decisions to an investment manager and be part of the decision-making.

It also offers estate and trust services, as well as a range of insurance products to meet physicians and physicians' families' needs. They provide continuing medical education and seminars. MD Management offers a range of services starting at the medical student level, and continuing through residency training, the establishment of your practice, the investment planning during your career and finally planning for your retirement. Consult the Canada MD Management website to review all the services they provide at every step of your career (*mdm.ca*).

You will be able to review the types of funds and mutual funds they offer and the annual rate of return. This provides a sample of the return you may earn on your investments with MD Management. You should certainly have all this information available before meeting with an MD Management financial advisor. Compare their results to other banking/investment institutions. The final decision is yours.

REFERENCE

1. https://mdm.ca/index.asp

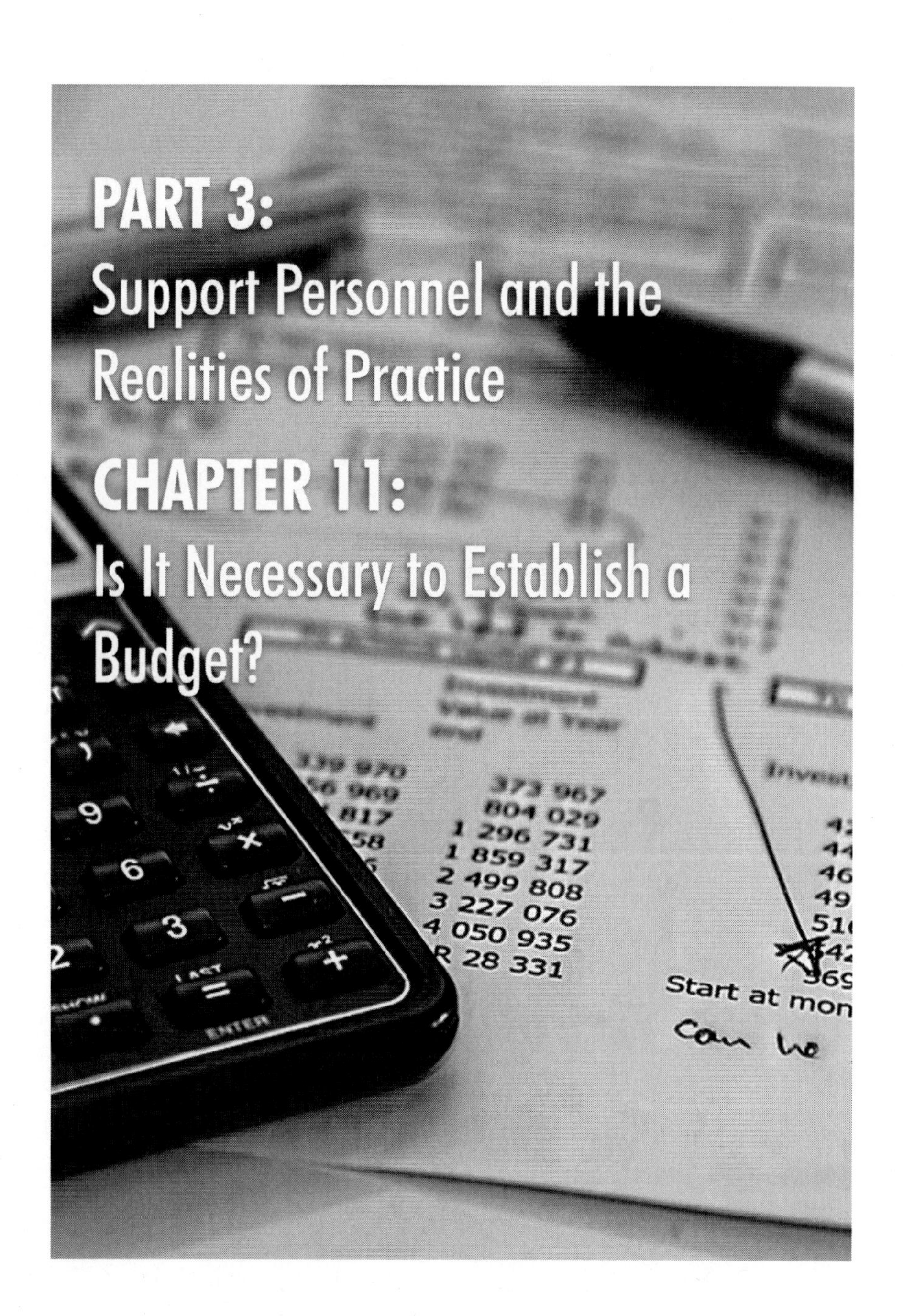

PART 3:
Support Personnel and the Realities of Practice

CHAPTER 11:
Is It Necessary to Establish a Budget?

Creating a budget may not sound like the most exciting thing in the world, but it is vital aspect of keeping your financial house in order. The budget will only be as good as the amount of detailed information that you gather. Ultimately, the budget will help you to understand where the money comes from, what the total income is, and what are the expenses. After reviewing your budget, you can see your net income and how successful your practice is.

Principles to Establish a Budget

1. Gather all your statements, expenses and keep the last few food, gas, entertainment and clothing bills, to calculate the average amount of your expenses.
2. Record all of your sources of income. If you are in a private practice, like most physicians in Canada, you will need to record your account receivables. You will also need to include your spouse's income if she/he works.
3. Create a list of the monthly expenses for your household and your practice. Write down a list of all your expenses you have had for both household and practice including house insurance, car insurance, house taxes and any other expenses you incurred during the year (expenses that occur during the year but not on a monthly basis) and average them out over 12 months.
4. Break expenses into two categories: fixed and variable. Fixed expenses are those that stay the same each month, such as a mortgage or rent, car payments, cable/internet service, house and car insurance yearly divided by 12 months and income tax monthly payments. These expenses are unlikely to change from month to month. Variable expenses are everything else, like food, wine, entertainment, gas, electricity, clothes, travelling, gifts, etc. It is important to be as detailed as possible for this category.

5. Calculate your total monthly income from your practice, your spouse and any other sources. As a physician, earning upwards of 6 figures and more if your spouse brings in an income, the end result should show a positive cash flow which you can plan to divert toward an RRSP, your investment retirement plan and your children's education fund. If, unfortunately, your budget shows a negative cash flow balance, you need to seriously review your expenses and make some hard decisions.
6. When you start to keep a budget, review your budget every 6 months at first and then, when you feel comfortable with your budget, you can review your budget on an annual basis.

It does not take that long to put all your information together on a spreadsheet. It is an exercise you need to do probably once, perhaps twice, and you will get comfortable with the status of your budget. The earlier you do this exercise in your practice, the better it will be for you to understand how much revenue you are receiving and how much money you are spending. This will provide you with a clear picture of your average net income.

Monthly Budget - Detail

Description	Category	Projected Cost	Actual Cost	Difference	Actual Cost Ranking
Airfare	Travel	$100	$0 ▲	$100	
Bus/Taxi fare	Transportation	$100	$150 ▼	($50)	
Car	Transportation	$50	$28 ▲	$22	
Car Insurance	Insurance	$500	$30 ▲	$470	
Charity 1	Gifts and Charity	$200	$200	$0	
Charity 2	Gifts and Charity	$500	$500	$0	
CMPA	Insurance			$0	
Conference Fees	Practice Requirements			$0	
Credit Card 1	Loans			$0	
Credit Card 2	Loans			$0	
Dining Out	Food	$1,000	$1,200 ▼	($200)	
Disability Insurance	Insurance	$0	$40 ▼	($40)	
Electric	Housing	$45	$40 ▲	$5	
Federal	Taxes			$0	
Fuel	Transportation	$450	$400 ▲	$50	
Health	Insurance	$400	$400	$0	
Hotel	Travel			$0	
Internet	Housing	$100	$100	$0	
Investment account	Savings or Investments			$0	
Medical Journal Subscriptions	Practice Requirements			$0	
Mortgage or Rent	Housing	$700	$700	$0	
Office Insurance	Insurance	$400	$400	$0	
Office Rent	Housing			$0	
Professional Dues	Practice Requirements			$0	
Professional memberships	Practice Requirements			$0	
Provincial	Taxes			$0	
Retirement account	Savings or Investments			$0	
Secretarial Salary	Salaries			$0	

When a person earns a high net income, it can be easy to feel that there will be plenty of money to pay the bills. More often than not, money is spent faster than it is being earned. It is easy to not be aware of how much we spend in a day or a week. It can be astonishing to see the amount of money spent on coffee, gas, food, restaurants, magazines, etc.

Budget Categories	Values Total Cost	% of Expenses
Food	**$1,200**	**28.65%**
Dining Out	$1,200	28.65%
Insurance	**$870**	**20.77%**
Health	$400	9.55%
Car Insurance	$30	0.72%
CMPA		0.00%
Disability Insurance	$40	0.96%
Office Insurance	$400	9.55%
Housing	**$840**	**20.06%**
Mortgage or Rent	$700	16.71%
Electric	$40	0.96%
Internet	$100	2.39%
Office Rent		0.00%
Gifts and Charity	**$700**	**16.71%**
Charity 1	$200	4.78%
Charity 2	$500	11.94%
Transportation	**$578**	**13.80%**
Fuel	$400	9.55%
Bus/Taxi fare	$150	3.58%
Car	$28	0.67%
Taxes		**0.00%**
Federal		0.00%
Provincial		0.00%
Travel	**$0**	**0.00%**
Airfare	$0	0.00%
Hotel		0.00%
Practice Requirements		**0.00%**
Conference Fees		0.00%
Medical Journal Subscriptions		0.00%
Professional Dues		0.00%
Professional memberships		0.00%
Salaries		**0.00%**
Secretarial Salary		0.00%
Loans		**0.00%**
Credit Card 1		0.00%
Credit Card 2		0.00%
Savings or Investments		**0.00%**
Investment account		0.00%
Retirement account		0.00%
Grand Total	**$4,188**	**100.00%**

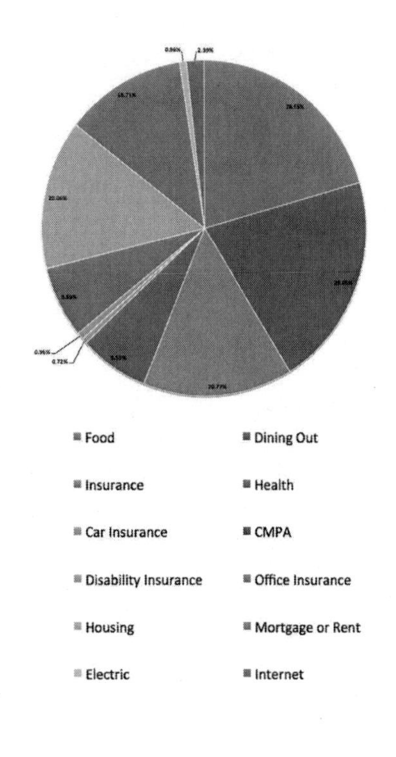

The most important reason to do a budget-spread sheet is to know where your money comes from, the total amount of money you are earning and where the money is spent, in detail. You want to understand the whole picture.

Budget Summary

Projected Monthly Income		Projected Monthly Expenses
Practice Income	$6,000	$4,545
Stipends	$1,000	
Administrative income	$2,500	
University Income	**$9,500**	
Investment/Rental Income		
Actual Monthly Income		**Actual Monthly Expenses**
Practice Income	$5,800	$4,188
Stipends	$2,000	
Administrative Income	$1,500	
University Income	**$9,300**	
Investment/Rental Income		
Balance (income - expenses)		
Projected Balance	**$4,955**	
Actual Balance	**$5,112**	
Difference	**($157)**	

Again, after a careful assessment of your expenses and your income, if you have a negative balance of income, you will need to review your budget and decide what needs to be done to bring your budget back to a positive position. Keep in mind that to balance your budget does not mean to get a loan to bring your budget back into balance. The budget should be balanced with income and expenses on a monthly basis without loans or relying on credit.

Budget and Spouse

A word of advice concerning marriage or having a partner. You may have been in practice for a while, accumulated assets and real estate. It is certainly appropriate to at least review the need for a pre-nuptial agreement. This isn't something to be sensitive about for either party; it is just a healthy precaution.

Canadian laws state that when you are married or have a common-law partner, the assets will be divided in half. Only in rare circumstances will it be different. But your net worth could be cut by more than half if, in addition to legal expenses, you may be forced to sell your vacation home, your own home, and potentially other investments to raise cash. You will also be paying capital gains tax.

For these reasons, a pre-nuptial agreement is advisable for a second marriage. You will need to protect the assets you have accumulated.

Do not let passion and love overwhelm your judgment. Your lawyer should be able to provide you with much-needed information about writing a suitable prenuptial agreement.

Concurrently, a divorce may occur. It is advised that you get the best divorce lawyer recommended by your own lawyer or through your own research. He/she may be expensive, but you must remember that you are in a situation that will affect the rest of your life. A good attorney should have a will document drawn up as soon as your divorce is finalized, again to protect the distribution of your asset as you wish.

Strategies to Manage the Positive Cash Flow

In contrast to this, let's say you are very careful with your money and you have a positive income balance every year. What should you do with this extra money, how should you invest it? What is the best way to maximize your returns? Irrespective of how much money may be available for investment, it is suggested that you create a formula for your finances. I have adhered to this throughout my career and had a great deal of success using this method:

a. You and your spouse (if he/she works) need to establish an RRSP (registered retirement saving plan) as soon as possible in your career. You may start during your residency training, even if it is just a little money. It is an important investment you cannot afford to pass by. You can put aside upwards of $22,000 per year into your RRSP's (your accountant can advise you on the exact amount you are eligible for in relation to your income). The same formula applies for your spouse. This money is deductible from your personal income tax, as well as your spouse's income. You can then invest this money as you wish. If you follow this rule and contribute to the maximum eligible amount of money every year for 30 years and invest this increasing amount of money conservatively, by any actuarial calculation, you will have accumulated an asset of $1.4 to $1.6 million over the 30 years' period; that is the period between 30-35 to 60-65 years old. Another word of advice is how to accumulate or save $22,000 or more each year. Do not wait for the banker or financial adviser to call you in January to tell you that you have not made your RRSP contribution. Most likely, you do not have the money (as I have seen many times) and you get the bank to advance the money on your line of credit, thinking you will pay it back soon. By doing this, you will negate all the benefits and advantages of saving and investing money. I personally suggest that you have your banker transfer $2,000 per month automatically out of your personal account and the appropriate account for your spouse out of his/her account to the financial advisor managing your investment account. You can interpret this as a monthly payment and include it in your budget-spread sheet. At the end of the year, you will have the money necessary to put towards your RRSP and your spouse's RRSP. This is a simple and effective rule to follow. This section of contributing to an RRSP applies to any physicians in practice, including family practice, specialty practice, academic practice, hospital practice and private practice. Be organized and plan ahead, it will pay off for you in the future.

b. You have now have a plan to save money toward your and your spouse's RRSP contributions. You need to develop a Savings Plan for your retirement. This section still applies to physicians in any type of practice arrangement, but the amount to put aside may vary according to your pension plan if you have one. In the chapter on your accountant, I will discuss the benefits of having your practice incorporated if you are in private practice or as a consultant practice. Being incorporated provides you with another vehicle to save money. Above the RRSP money you have contributed in the year, it is a very good banking practice to open a personal savings account, a practice/company savings account and a children's education savings account.

Refer to your budget spread sheet, as it will provide the information on how much money is left after expenses in the household column and in the office/practice column, providing the cash flow you need to invest toward your retirement and your children's education. Then you can establish the annual amount divided by 12, directed monthly by your banker to each savings account. This money can be sent to your financial advisor at the end of the year for long-term investments. This plan and amount of money can be reassessed as often as you want with your life style and as your income changes. Again, you do not have to follow this plan and schedule, but I can attest to the soundness of this regimen.

The bottom line is, it is easier to spend money than save money. You should take all the steps possible to follow a money-saving strategy and adhere to it. You can start to save money when you are a resident and continue this habit of savings through your entire medical practice, which will result in financial security. Financial security is an incredibly powerful asset.

PART 3:
Support Personnel and the Realities of Practice

CHAPTER 12:
The Use of a Lawyer

Almost everything we do, from purchasing a home to driving a car or interacting with others, is affected by the law.

Why?

The need for assistance in legal-related issues can be difficult. There are an overwhelming variety of choices. There are multiple subspecialties in law and you may have the challenge of having to choose the right type of lawyer for your needs. Therefore, you have to differentiate the specific issues you are dealing with and find the appropriate lawyer. But establishing a relationship with a particular lawyer will be very helpful.

A lawyer will help you with buying a house, purchasing real estate for your practice or your family, starting your practice, engaging in a partnership, advising you on your contract with any organization you will work for or be affiliated with, counselling you on tax matters, establishing your estate planning, will, divorce, etc. It is indeed a very good idea to have an established relationship with a lawyer that you will work with. He or she can help you with any of the above-mentioned possibilities and in the event that you come across an unusual legal situation.

Who?

A trusted lawyer will be there to take care of the issues you need help with and/or will refer you to the right person who can provide you with assistance, such as in the case of a complex tax issue (dealing with Canada or another country tax services), in the case of a complex and difficult divorce, or in medico-legal cases, discussed later in this chapter.

The process of finding a good lawyer that you trust and can develop a professional long-term relationship with starts with just a few basic steps:

1. Shop around and ask questions.
2. Get recommendations from friends, colleagues, and family.
3. Determine the issues of why you would need a lawyer so you can find the appropriate help.
4. Find a lawyer who will evoke feelings of trust and confidence.
5. Choose a lawyer who is upfront with his/her fee structure and their fees for different services.
6. Check the lawyer's credentials and his/her reputation.

Informed Consent

The basics of informed consent should include the following:

1. Discussion with the patient regarding the diagnosis. If there is some uncertainty about the diagnosis, mention the reason for it and what else is being considered.
2. Discussion with the patient, the proposed treatment and its complications, including any special risks, perhaps small, relating to the treatment. If the treatment carries serious consequences — such as death, for example — it must be disclosed.
3. Answer any specific questions posed by the patient. Be alert to a patient's concerns about the proposed treatment and address them appropriately. The patient must always be given the opportunity to ask questions about the risks involved in the proposed treatment.
4. Inform the patient about the consequences of leaving the illness untreated. Although you should not be forcing the patients who refuse treatment, you have an obligation to inform patients about the potential consequences of their refusal.

5. Pay special attention in obtaining consent for optional or cosmetic procedures. When obtaining consent for cosmetic surgical procedures, or for any type of medical or surgical work that may be less than entirely necessary to the physical health of the patient, you must take care in fully explaining the risks and anticipated results. If the treatment is part of experimental research, extra care should be taken to ensure the patient understands the purpose of the experimental study. At the very least, the patient should read the study summary and sign a special research consent form.
6. Inform the patient about reasonable alternative forms of treatment and their risks and benefits. There is no obligation to discuss what might be clearly regarded as unconventional therapy, but patients should be made aware of other accepted alternatives and why the recommended therapy has been offered.
7. Although a patient may wave aside all explanations, ask no questions, and be prepared to submit to the treatment whatever the risks may be without any explanatory discussion, you must continue to exercise your obligation to provide sufficient information for informed consent.
8. Do not guarantee results. State the exact information about risks and provide encouragement about optimistic results of treatment. Do not allow for misinterpretation by the patient that results are guaranteed.
9. You must be a clear about the need to proceed at the time. In medical emergency situations, treatments should be limited to prevent prolonged suffering or to deal with imminent threats to life, limb or health.
10. Even when the patient is unable to communicate in medical emergency situations, his or her known wishes must be respected. Before proceeding, the physician will want to be satisfied there has been no past indication, by way of an advance directive or otherwise, that the patient does not want the proposed treatment. Further, as soon as the patient is able

to make decisions and regains the ability to give consent, a proper and informed consent must then be obtained for additional treatment. In many provinces, legislation permits the designation of a substitute decision-maker to provide or refuse consent on behalf of the incapacitated patient. If the substitute decision-maker is immediately available, emergency treatment should proceed only with the consent of that individual.

11. In urgent situations, it may be necessary or appropriate to initiate emergency treatment while steps are taken to obtain the informed consent of the patient or the substitute decision-maker, or to determine the availability of advance directives. However, the instructions as to whether to proceed or not must be obtained as quickly as practicably possible.
12. When an emergency dictates the need to proceed without valid consent from the patient or the substitute decision-maker, proper documentation explaining the circumstances necessitating the physician's action is essential. If the circumstances are such that the degree of urgency might be questioned at a later date, arranging a second medical opinion would be prudent.
13. When the patient (or substitute decision-maker) is unable to consent and there is demonstrable severe suffering or an imminent threat to the life or health of the patient, a physician has the duty to do what appears immediately necessary without consent. Emergency treatments should be limited to those necessary to prevent prolonged suffering or to deal with imminent threats to life, limb or health. Even when the patient is unable to communicate, his/her known wishes of must be respected.
14. Where part or all of the treatment is to be delegated, patients need to know that others will be involved in their care. Consent explanations should include such information.
15. Write it down! A written note documenting the consent discussion can later serve as important confirmation that a

patient was appropriately informed, particularly if the note refers to any special points that may have been raised in the discussion.

Tips for Dealing with Adverse Events

1. Deal with any emergencies and immediate health concerns.
2. Residents involved in an adverse event should report it to their supervising physicians and are encouraged to be present to observe the disclosure discussion as a learning experience. If time allows, Canadian Medical Protective Association (CMPA) members may wish to seek telephone advice from the association prior to communicating with the patient, family or hospital involved.
3. Give your patient factual clinical information about what has happened and the clinical nature of his/her condition as it now exists. Avoid speculation about what may have happened if a different course of action had been followed. Avoid attribution of blame, particularly concerning the care provided by others.
4. Provide recommendations to deal with the medical condition, as it now exists, including alternate treatments and the risks and benefits of any other investigations and treatments. This is an informed consent discussion on how to move forward. Answer your patient's questions about the proposed treatments.
5. Maintain close communication with your patient and the family (with the patient's consent) about the ongoing clinical condition and any further plans for treatment.
6. Facilitate any necessary treatments and consultations.
7. Transfer the care to another physician if your patient requests or prefers it, or if the condition requires care that you cannot provide.

8. Document your care and the discussions that occurred in a factual way after the adverse event.
9. Never alter the record or change what had been written previously in any way.
10. Call the CMPA if you are concerned about a medico-legal problem as a result of the incident/complications.
11. Express to the patient and family your feelings of empathy, sorrow, and concern as appropriate. Sharing your sincere regret about what has happened, or wishes that the event had not occurred, is an entirely acceptable and desirable response. Sometimes, if the outcome is indisputably due to your improper care, you may acknowledge your responsibility.
12. Inform your patient about any process through which the incident may be investigated, but be aware that there may be limitations on what information may be made available from further analysis.

Canadian Medical Protective Association

The Canadian Medical Protective Association (CMPA) is an organization based in Ottawa, Ontario. It provides legal defense, liability protection and risk management education for physicians in Canada. It also provides compensation to patients and their families when they have been harmed by negligent care.

The CMPA was founded in 1901 and most Canadian Physicians are members and have medical coverage from CMPA. It is an easy decision to make as you start your practice and need medical liability protection; CMPA provides full coverage except if you opt out, requiring then a private insurer.

The CMPA has been credited with helping to control health care costs in Canada by enabling physicians to practice good standards medicine and avoid duplicate and unnecessary tests. The CMPA also helps avoid complex extensive documentation as one may see in the American health care system, where the medico legal aspect of practicing medicine is very strong.

With this approach, the cost of medical malpractice in Canada is around one-tenth of that in United States, notwithstanding the difficulty in finding proper private malpractice insurance coverage at a reasonable cost.

CMPA is funded partially by annual Physicians' fees and partially by the government to a ratio around 20%/80%. CMPA has in recent years reviewed its services, which continue to provide safer medical care and reduce the number and severity of medical adverse events. To ensure this, CMPA hosts risk management conferences and symposiums for Canadian Physicians, delivers around 400 customized workshops, and publishes a quarterly magazine (CMPA Perspective).

CMPA partners with the Canadian Safety Institute to develop programs and courses to help physicians with patient care and safety. The list of programs is available on the CMPA website.

CMPA funds the legal defense of Canadian Physicians when the case is in civil court for medical negligence causing injury or in criminal court for offenses ranging from financial fraud to sexual misconduct, drug issues and felony crimes. In defending the Canadian Physicians, CMPA works hard to settle cases and pay out the lowest compensation possible. On the CMPA website in the references you can view recent information on the number of legal actions in different stages.

With Internet access, patients can obtain objective medical information and patients are more demanding in medical transparency and accountability. It is certainly helping to motivate doctors to keep up to date and learn new medical information and technology.

The medical malpractice definition is professional negligence by act or omission by a physician in which the treatment provided fell below the accepted standards of practice and caused injury or death to the patient. If a physician is served with a legal notice from a lawyer on behalf of a patient, the first step for the physician is to call CMPA and speak to an officer about the received document.

A CMPA file will be opened and the officer will direct the physician to a CMPA lawyer representative in the same town or the nearest office. As physicians are usually not knowledgeable in medico legal matters, it is a good idea to read the information on medical practice provided by the CMPA website.

The website will provide background information on medico legal procedures, the parties and their role, the elements that constitute a case, the process of filing, examination for discovery, expert testimony and trial.

Discuss with your CMPA lawyer the potential consequences for your practice and/or your file at the Royal College of the province where you are practicing.

I had the unfortunate experience at the beginning of my practice of a medico legal issue when the other party was "leaking information" to the College. It was unethical and unprofessional but damage was done and the College requested information that otherwise in normal procedure would not have been needed. Such behavior is highly unlikely but one can never be too prepared.

Hopefully, it will be rare you will ever need legal counsel from CMPA for your practice but if you do, follow the steps recommended by CMPA, establish a good rapport with the CMPA-designated lawyer. Be forthcoming with the information, as this is to your benefit. Regardless of whether you believe you are right or not, inform CMPA whatever you felt you did or did not do regarding negligence, an error or a mistake.

You can do a lot to avoid malpractice claims or other complaints about your medical care. The most controllable factor is the quality of your relationship with your patients. In many cases, it is the physicians' attitude that is the chief cause for action against the physician. Indifference, arrogance or anger can trigger such a lawsuit.

When a treatment or procedure has resulted in an adverse outcome, while the doctor's first impulse may be to avoid a discussion, the physician must still discuss with their patients the outcomes of care that differ significantly from anticipated outcomes.

It is a simple step in preventing a lawsuit, so do not delay. Silence may be interpreted by the patient or family that the physician has something to hide. Select a private place for the discussion. A hospital hallway or waiting room is disrespectful. Express compassion and care and avoid defensiveness. The patient and family may be upset, vent their anger and challenge the physician who knows better than to react. Be honest, explain what happened and admit errors/mistakes if there were any. Do not blame others even if you are tempted as it does not really help the situation and it does not necessarily protect you from a lawsuit. Apologize if needed: the family will appreciate this gesture.

Guidelines when Testifying

1. Always tell the truth, even if it means talking about your problems/errors/mistakes. Sincerity and honesty are your best options.
2. If you do not know the answer, admit that you do not know and do not try to create an answer.
3. Be sure that you understand the question and let the person finish before you respond. If you do not understand, the question will be repeated.
4. Answer the question that is being asked and then stop. Do not open the door for difficult or further questions by adding comments even if you think the comments could help.
5. Answer directly to the person asking the questions, except if your lawyer interrupts you.
6. Speak clearly and slowly so people can understand you. This also gives you time to compose your answer.
7. Do not argue with the lawyer of the other party because you may think you know better or that he/she is wrong. Do not be sarcastic or try jokes. This is not viewed well.

REFERENCES

1. (en.wikipedia.org/wiki/Canadian_medical_protective_association)
2. (en.wikipedia.org/wiki/Medical_Malpractice)
3. https:/www.cmpa.acpm.ca/

PART 3: Support Personnel and the Realities of Practice

CHAPTER 13: The Big Word: Investing

At some point after you have started your practice, you may find you have begun a comfortable lifestyle and have some disposable cash on a more regular basis. This is a great time to develop an investment program: start an RRSP if have not already done so, consolidate your and your spouse's student loans, and establish a registered education fund (RESP) if you have children or when you have children.

The more you read and the more you study, the more you learn. It is the same with investing. For the young investor just starting out, who has usually very little or no investment knowledge, it can feel overwhelming trying to understand the amount of information available. The list in Addendum 1 features interesting websites that you can review and learn more about investment strategies.

Why Investing?

There are two components of a successful investment program. The first is making sure that a fixed amount or percentage of your income is saved regularly. This is purely a matter of self-discipline. You should have a system for personal savings and regular contributions to your children's savings funds that are regularly taken from your accounts. Electronic banking systems make this as easy as automatically paying a bill.

The second part of a successful investment plan is to decide how savings will be invested. Investing means putting your money to work for you. You can earn an excellent income as a physician but there is a limit to how many hours a day one can work, so investing your money will increase your assets over time. There are many different ways you can go about making an investment such as certificates of deposit, bonds, stocks, mutual funds, hedge funds, derivatives, other investment asset classes such as precious metals, real estate and more. Each of these has positive and negative aspects and as nothing is perfect, whatever method you choose for investing your money you should meet your long-term goals.

Investing should not be gambling and you should not put your money at risk. Working with your financial advisor and reviewing the wide array of financial vehicles, their market strengths and weaknesses, will help you to invest your money toward long-term financial security. For most physicians, investing is a necessity, as most doctors will not have a pension fund/plan when they retire, and they will rely on their investment income to maintain a certain lifestyle.

Investing Principles

1. Build an emergency fund. Set aside at least 3 to 6 months' living expenses as an emergency fund. It should be money readily available so you do not have to sell something and pay capital gain or lose money in selling the assets.
2. Before investing money, pay off your debts. If one loan is considerably smaller, pay it first. Pay off the loan with the highest interest rate as fast as you can. If one of the loans interest rate is deductible, pay the non-deductible loans first. Student loans do not need to get paid until you start working and the interest rate may be quite low. You may be able to pay it off completely when you have extra cash. Otherwise, use the return on your investments to pay the interest on these lower-rate loans.
3. As stated in Chapter 11, set up your financial goals for your short-term and long-term investments. While you may not need a financial advisor in order to invest, it is still very sensible for a busy physician to have a professional who understands market trends, studies investment strategy and can diversify and manage your portfolio.
4. Generally with investments, the riskier, the higher the potential for return on investment, but also the greater potential to lose your investment. Very low risk investments

such as bonds and deposit certificates generate a very low return but at very little risk.

5. A portfolio should be diversified to spread the risk. In general, a person could have 70% in equity, 20% in fixed income and 10% in cash. You can modify the ratio according to your own plans. Diversification spreads the risk, maximizes your return and can help weathering the ups and downs of the market.
6. Invest in companies and sectors of the economy that you understand and that will help you meet your goals. Choosing an investment just because a certain stock is gaining or because a friend told you about an "amazing" stock is not a good strategy.
7. Whatever you invest in, invest for the long run. One may see the stock market as an opportunity to make a quick dollar in a short time, but it is usually the opposite. Day trading is even more risky because you need time to do it, to understand it, to deal with the unpredictability of the market and pay the fees. So invest for the long run and for years to come.
8. Many friends, many investors, and many books will tell you to buy low and sell high. We all know it makes sense, but it is not that easy. When a stock is cheap, there is always a reason. Stocks typically drop in price because there is a problem with the company or with the product, or the market. When the entire market drops like the big sell-off of 2008 or during a "significant correction", it is possible to find certain stocks that fall simply because of an overall negative sentiment. That is when a long-term investor can find great long-term value positions.
9. Weather the storms. You may be tempted to sell when you see the value of your investments plummeting. There are always swings in the market. When there is a bear market, assess what happened and make sure that the fundamentals in the stocks you own are good. You may even want to buy more stocks at a lower price if the fundamentals are good. If not,

then you may consider selling. Losses can always be offset against other capital gains.

Investing Options

a. Invest in a savings account. It offers low return but you do have the cash available in case of an emergency. Money market accounts can earn a slightly higher interest rate and can also be accessed when you need the money.
b. Save with deposit certificates as part of your fixed income strategy. You can invest on a short term or long term with a higher yield. Another fixed income strategy is bonds. A bond is basically debt assumed by a government or a company to be paid back later with interest, regardless of the market conditions. You do need the help of a financial advisor to navigate the world of the multiple bonds available.
c. Stocks: Stocks are purchased from your financial advisor or on-line broker. In buying a company stock you buy shares of a public company, which entitles you to receive a fraction of the profits that are paid as dividends. Over a long period of time, large capitalized stocks are safe and usually profitable. As there are thousands of stocks world-wide to choose from, it can be a headache so make sure your criteria works for you. As a general principle, I would advise you to choose bank stocks, commodity stocks, mining stocks, medical/pharmaceutical stocks and blue chip stocks. Diversification is a good rule to follow as it decreases your risk across multiple stocks if one goes down. You should have some idea about each stock you are buying. Of course, there are multiple ways and formulas to assess a stock. Though they are complex, they are not a certainty as to which relates to the stock's future performance. Remember to always look at the past performance of a stock or a company when you make your assessment although that does not always represent what the future may or will be. I

believe it is the job of the financial advisor to discuss in detail your ideas and your plans and review with you any vehicles of investment that might suit your needs and objectives. Trust is a must. Doing your own investment and trading as 'day-trading' will use important time in your day which can be used to see patients or be with your family. It is also not what you have been trained to do. As well, it also increases the level of stress because realistically, you are nervous about the decisions you make unless you are very comfortable with the complexity of the market, and I am sure it is not easy to achieve that, as it is not your career.

d. Mutual Funds: Mutual funds are managed portfolios of stocks and bonds bundled together. There are thousands of mutual funds to choose from. Mutual funds should be a long–term investment. Your financial advisor should help you with historical performance, data tracking, and performance consistency in relation to your goals. Remember that, again, historical performance is no indication of how the fund will perform in the future. In general, mutual funds are long-term investments, but they should be reviewed regularly as with the rest of your portfolio.

e. Precious Metals: Gold has been a money asset and currency of trade since the beginning of civilization. Its amount is limited pending new discoveries in the future and 85% of the gold available is above ground in the hands of banks, agencies and people. Owning gold can be part of one's portfolio or it can be solid gold placed in your bank safe and the return can be realized only when it is sold. The purchase of gold coins in small quantity is also possible and they are easier to re-sell. Usually exposure to gold is through ownership in gold stocks. These stocks can also provide the added benefit of dividends. You can also buy precious metal mutual funds and Exchanged Traded Funds (ETF's). There are many ways to invest in precious metals but it is best that you consult your financial advisor.

f. Commodities: You cannot live without commodities: food, fuel, electricity, and water. It is a very specialized area of the market and there are many options to invest in. I would recommend being very careful: get some understanding of commodities and work with your financial advisor. This will allow you to choose the type of commodities that work for your planning. Again, commodities should be held over a long term and reviewed every year, as they do swing with the volatility of the market and not necessarily because they are a poor choice.
g. Collectibles: Collectibles are fun and enjoyable but few people realize a reasonable return from the money spent on collectibles. There are a few points to remember about collectibles:

    1. A collection must be focused on a specific object/item.
    2. You must be knowledgeable about what you want to collect or are collecting.
    3. You have to buy quality at the most acceptable price.
    4. You must start small, as your tastes will change over time.
    5. Patience is required as it takes a long time to build a collection.
    6. Forgeries abound, so choose from reputable people or agencies.
    7. Talk with people as not every dealer has the same opinion.
    8. It is very hard to predict the future value so if your investment does not materialize, you still have a collection.
    9. Collections are not always easy to sell and there can be a commission.

Buying art is very similar to collecting but art is there mainly to enjoy. Don't buy just to buy but buy because you like it. Do not buy at auction events like dinner or golf tournaments, as you will not get the best value for your money. Give a money donation instead, as the whole amount will be deductible.

h. Selling Short: Everyone knows how to profit in the long-term: buy low, hope the price goes up, and then sell high. Selling short is the reverse. Sell high, hope the price goes down, and then buy back at a lower price. It is a basic technique of investing. It may help in controlling some of the volatility in the market and it is another tool to make money and preserve your capital. There are some negatives to this approach, such as the possibility of losses because you can easily miss the market opportunity and short-selling requires exceedingly close monitoring of the positions you are short selling.

i. Hedging and Hedge Funds: these are complicated managed portfolios that use market derivatives and can be risky. Generally, only do this once your portfolio is secure and you can afford to take some risk.

j. Venture Capital: People may have great ideas but need financing to start the project, buy equipment, hire people, perform research and help market their ideas before they earn any money and make a profit.

   Venture capitalists provide assistance and raise the financing required to help people start their project/business in exchange for some amount of the equity. The principal downside of investing in venture capital is that you may not get your money back but you may profit if the project/business gets established and earns a profit; otherwise, you may have to wait a long time while you get no return on your investment. If you have invested in a venture capital and hold shares, the venture capitalists may raise more money, issuing more shares therefore diluting the purchase price of your shares. If you want to sell, it is difficult because it is not traded and most likely you will sell at a loss. Out of 100 venture capital projects, very few will get established, make profits and perhaps one day be traded.

k. Derivatives: Derivatives are complex financial tools used by high-power investors. They are risky, and everybody remembers the 2008 financial collapse. This was mainly caused by the explosion of derivatives. These products are artificially developed, their real value is not always known, and they are still a risky investment for any portfolio.
l. Futures Options: These are derivatives that are primarily used to hedge commodity portfolios for commodity producers. Usually it is speculators who trade future options, as it is a complex process.
m. Real Estate: Your home is generally a good investment as long as it is not overly expansive and that you do not have a large mortgage with payments that will make you what we call "house poor". It is a false perception to think you are worth the new assessment of your house if you have a large monthly payment and in Canada, mortgage payments are not tax deductible.

Amortizing your mortgage, large or not, will represent a large amount of money paid in interest over 20 or 30 years. You should buy a nice house, a house appropriate for your family, and have the smallest mortgage possible. It is important to ensure that your monthly mortgage payments do not include just the interest but also some principal, which helps you pay the balance of the loan.

Home equity is great but also dangerous. You should not use your home equity as a line of credit to buy other things like a car, a boat or a trip. Ultimately you just increase your debt. In general, the less expensive the house, the more marketable it is the more expensive, the less marketable. And remember the sale commission, as it can be substantial. It is also not advisable that you sell your house privately and try to save the commission. It is not easy to advertise, to show the house and to process the papers. A real estate agent is worth his or her commission.

Try to live as close to work as possible, 20 to 30 minutes maximum. Spending 45 to 60 minutes in your car each way is not a good use of your time as it is stressful in traffic and bad weather. As much as possible, buy a smaller home reasonably close to work and perhaps invest in a recreational property a few hours away. Buying a home is an emotional time and the real estate agent knows it. Assess carefully if the house you are interested in meets your family needs, if you like the environment, and if you are close to schools and to work. Try not to capitulate to the pressure of the real estate agent's selling points, such as "There is a need to make a speedy decision", "There are other offers", or "The house is a great value". Always remember, no deal at all is better than a bad deal, and there will always be another house. If the real estate agent gets impatient or difficult, he/she is not worth your time.

The most desirable way to do remodeling and/or make additions to the house is to pay cash or through a small loan that you can pay back in a short period of time. Remodeling or additions to your house should be consistent with the look and the value of the house so you can get your money back when you sell. It does not mean to do cheap renovation or additions, but try to complete renovations or additions that are not overly expensive and look good with the rest of the house. Unless you are extremely talented, you should not attempt to do the remodeling yourself. It takes time; there are construction codes you need to know (even more so if you want to do structural, electrical work and insulation, or plumbing). You will save time and money by hiring a proper contractor.

Another real estate option is to buy into the real estate of your practice such as the building you are in, the space you are in, or as a part of your practice. You can pay yourself rent (paying your own mortgage) as opposed to paying someone else's. You can also deduct your interest payments as expenses. The building, or the office space, remains under your control (maintenance, rental space). The building/space may increase in value over the years. If you buy with several colleagues, set up a proper partnership through your lawyer including a buy-out process by your colleagues if you want to sell and, conversely, buyout of your colleagues if they want to sell. Buying a piece of land and building a commercial building is feasible, but is a complex matter, with lots of decisions and larger risks; you need to consider it carefully.

If you are comfortable with the process of renting out, finding tenants, taking care of a property and assisting your tenants whenever needed can make it a good investment to buy a rental property such as a house but, probably easier, a condominium. It is good for the diversification of your portfolio. If your practice is incorporated, you can buy this rental property through your practice and deduct interest rates and expenses against your revenue. Over time, the property will increase in value and should provide be a good return when sold.

Awareness of Particular Situations

For one, you should avoid investing with in-laws. As a practical matter, if your marriage should end in divorce, they are no longer your relatives and may/will not side with you in any dispute. If an investment is made with a relative, it should be totally business-like with the advice of your lawyer. Detailed papers and loan agreements must be drawn up and signed with your lawyer for your own protection. As far as your children are concerned, they should be able to handle their own money and investments, but you may provide some direction. You should not support them in a business venture that may hurt you financially and jeopardize your retirement. It may be hard to discipline your children as a business owner if you disagree with their decision–making and the direction of the business.

Tax Shelters: there is no such thing as a true tax shelter. There are ways to delay paying taxes and minimize the amount of taxes, but there is no way to avoid taxes. You earn money, you pay taxes. Do not listen to friends or colleagues who claim they avoid taxes. High quality tax advice is always worth the fees paid to avoid an audit.

International Investing

Although most companies are already international, you need to know the international market you are interested in, the currency, and the rules. Foreign countries can change rules and seize assets, including the one you may be invested in. Our Canadian and U.S. financial institutions are efficient, well-regulated and provide access to the world market through strategies such as mutual funds and others.

Summary of Investment Principles

1. Start saving early. Time is money and you can at least start saving money in an RRSP, even during your training.
2. Find a reliable financial advisor and establish a long-term relationship with the advisor built on trust and understanding
3. Keep a steady course. The only road to wealth is regular savings and intelligent investment. Establish a home budget, a practice budget, a savings budget and stay the course.
4. Make sure to have some cash for bills and emergency situations.
5. Maximize your income tax options to save for your retirement and for your children's education.
6. Diversity reduces risk. Do not have all your eggs in the same basket.
7. Pay the least amount of fees possible, as money paid for these fees is money out of your pocket. When you do pay for fees, get the most value possible in solid advice.
8. Invest in proven investment strategies.

REFERENCES

1. R. M. Doroghazi
   The Physician's Guide to Investing: a practical approach to building wealth
   Humana Press, Springer Science, New York, 2009
2. R. M. Doroghazi
   Lack of Instruction on personal financial management and investing in medical education
   The American Journal of Cardiology, 98: 707-8, 2006
3. S. Orman
   The Road to Wealth: a comprehensive guide to your money
   Riverhead Books, New York, 2008
4. N. Rothery, D. Ashton, D. Bortolotti
   Guide to Investing in Stocks
   Rogers Publishing Ltd., Toronto, 2014
5. B. G. Malkiel
   The Random Walk Guide to Investing: ten rules for financial success
   W. H. Norton, New York, 2003
6. S. Kirschner, E. Mayer, L. Kessler
   The Investor's Guide to Hedge Funds
   John Wiley & Sons, Hoboken, NJ, 2006
7. G. Kleinman
   Trading Commodities and Financial Futures: a step-by-step guide to mastering the markets
   Financial Times Prentice Hall, 3rd Ed., Boston, 2005
8. M. C. Thomsell
   Getting Started in Options
   John Wiley & Sons, 7th Ed., Hoboken, NJ, 2007
9. M. El-Erian
   When Markets Collide: Investment Strategies for thE Age of Global Economic Change
   McGraw-Hill, NY, 2008

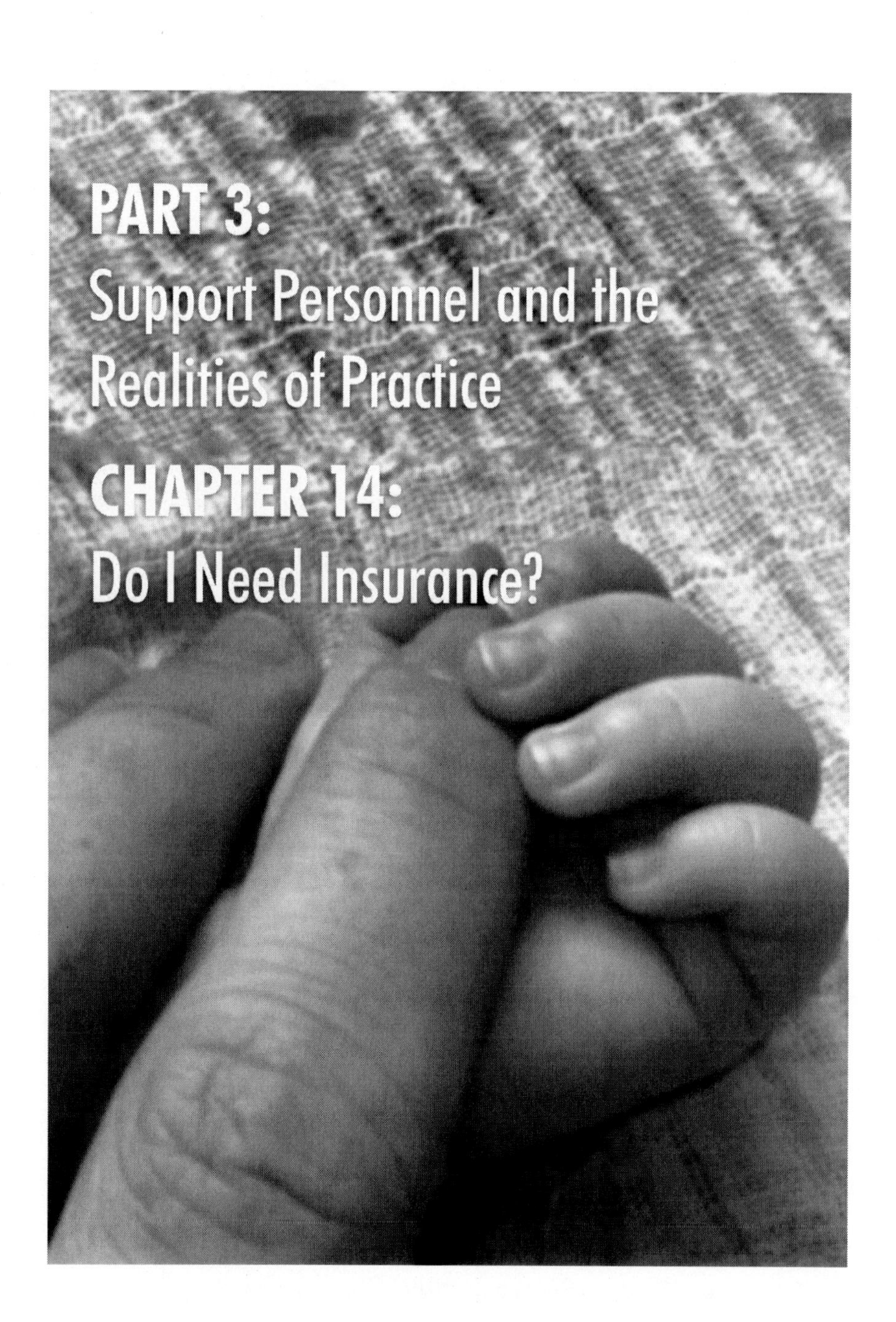

PART 3:
Support Personnel and the Realities of Practice

CHAPTER 14:
Do I Need Insurance?

At some point in life, we all need insurance. It could be car insurance, homeowner insurance, life insurance, disability insurance, professional insurance, estate insurance and so on. The main function of any insurance is to protect us against loss. This could be a loss of something that we cannot afford to lose, sustain or that we will need to replace with our insurance. Some examples of how insurance protects us from such losses are: loss of income, loss of home, loss of life and loss of personal belongings.

What is Insurance?

Insurance is not an investment in the sense of saving money or investing monthly. You do not get a yearly return on your insurance; on the contrary, you pay your policy fees on a monthly basis with no return unless you sustain a loss or make a claim. Insurance is not meant to enhance your spouse or children's lives, although many people may think this. I have colleagues who say "I have a great life insurance policy, my wife will have a lot of money if I die."

The real purpose of insurance is to cover the basic needs, pay the bills, and provide support. In trying to make your spouse or children rich, it is possible to be over-insured and you end up paying excessive fees and money that could be better invested elsewhere.

Car Insurance

One cannot drive without car insurance, but there are many types of insurance plans in different parts of Canada. Each province has its own car insurance plans and policies. Seven provinces that have a private insurance plan through private insurance companies are: Alberta, Ontario, New Brunswick, Nova Scotia, Prince Edward Island, Newfoundland and North West Territories.

In each of these provinces, it is mandatory by law for a person to buy a minimum amount of insurance coverage.

British Columbia, Saskatchewan, and Manitoba have established a government-run car insurance company and the citizens in these provinces need to use the delegated government plan *(http://www.icbc.com/)*.

The province of Quebec is different, as there is both a government plan and a private insurer plan. The government insurance plan insures against all injuries to people and the private insurance companies cover the property damage. There is a minimum amount of third party liability insurance of $50,000.

Irrespective of the province you reside in, a few principles are worth mentioning. Third party liability coverage should be $1M to $2M. Increasing it to more than $2M of coverage will cost you significantly more and you will receive little benefits for the extra coverage. This may be different in the USA.

Basic coverage is required and may vary from province to province. The collision deductible will vary across the country and will add to the cost of your insurance. As the costs increase, he cost will increase, the smaller the deductible you will have.

Replacement cost coverage and new vehicle replacement coverage is certainly not mandatory and will be a personnel choice. This is an important discussion to have with your car insurance agent. Some believe that if the car is older than 3 years, replacement coverage may not be necessary. It certainly makes more sense to have this coverage for the first few years of a new car. If you have a new car, you may want to check with the car company (if your car is leased) as there might be replacement coverage included in your lease. Finally, nothing replaces an excellent driving record. Having a safe driving record can amount to a substantial insurance discount.

Homeowner Insurance

Homeowner insurance is vital to have to protect your home. We all value our family home as an important asset, financially and emotionally. Homeowner insurance is also a requirement by mortgage companies. There are many plans available and breadth of coverage and limits of liability vary from policy to policy and insurer to insurer. The purpose of this book is not to detail all of them, but provide some general guidelines.

House insurance coverage is usually assessed in relation to the square footage of the house and the appraisal of the house. Most of policies will cover 90 to 100% of appraisal value and contents, including rebuilding cost up to full house replacement. One needs to add third party liability coverage, usually $1M coverage.

In some high-risk areas of natural disaster such as flood, landslide, hurricane, tornadoes or earthquake, appropriate additional insurance may be important and useful. You may want to discuss this with your agent. You will want to be careful that the amount of coverage keeps up with the inflation and the value of your house. Any coverage on the house or contents should be reviewed every 2-3 years. If your home has specific built-in features when you purchase the home or increases in value due to renovations, you may want to request a special clause to cover these features at comparable cost and quality.

The contents of a home are typically insured to a percentage of the total value, usually 50% but you may request a higher replacement value for a higher policy cost. Probably more important is the adequate coverage of particular valuables, such as jewelry, art/paintings, antiques, musical instruments and/or sport equipment. An excellent way to record all the valuable household/belongings is to photograph them on a mobile device and download the information to a computer. It is a good idea to back up the information to your computer and your insurance company. This detailed information will avoid any confusion in the event that you need to make a claim.

If you have multiple insurance policies with the same company, you may be eligible for a discount. It is good idea to do some research before signing multiple insurance policies with different companies. When you speak with the various insurance companies it is a good idea to ask what products they offer and you can decide if it meets your requirement.

Disability Insurance

Your most valuable asset is your skill as a physician or surgeon. Another important aspect to this is your earnings. Therefore, a disability insurance policy is meant to protect your 20 years (or more) of education. Disability insurance will be there in case you are not about to practice due to an accident or illness. What is the definition of disability? Is it a general definition of being able to perform any kind of work or does it relate to performing work in your particular specialty? In insurance language, this concept is referred to as the owner's occupation (as compared to any other occupation). The former definition means that one is disabled if they can no longer perform their current occupation (like surgery or anesthesia). The later means that you are disabled only if you cannot perform any work at all (for example, if you cannot do cardiac surgery, but you can teach or do research).

When considering the purchase of disability insurance, you may want to consider the "waiting period." The time before the policy starts to pay benefits, the benefit period and the time when the policy will stop paying the benefits. The more quickly the policy provides benefits, such as one or two months after the injury or trauma, the more expensive the insurance can be. The longer the period before the benefits starts (such as 3 months) the lower the premium. It is recommended to obtain the longer period option, as the average physician should have sufficient savings to sustain their own expenses for at least 3 months. Remember: it is a good idea to keep some cash for unplanned events. This may prove to be very helpful if you need it. The benefit period can be the length that you chose. This period is usually to a maximum of age 70.

There are general limits on the amount of disability coverage that may be purchased. Benefits are usually limited to approximately 2/3 of a person's income, so in the event that you have take out your disability coverage, you may need to make adjustments to your current lifestyle and habits. Multiple plans are available and you should discuss all possibilities with your agent. It is recommended that you purchase disability insurance as soon as you start your practice. It is less expensive when you are younger and you will have less risk factors.

Remember that firstly, the disability insurance will not cover your full income, and secondly, that your income will probably increase over the years. Your expenses over the years may increase as well. So you may want to review your policy every year and make sure that it continues to match your annual income. This will help you to ensure that you have sufficient coverage for your basic needs. Check if you have the option in your current policy to increase your coverage by (at a minimum) cost in 5 to 10 years. Buying a new policy at 50 years old can be very costly. Also, some policies offer a cost of living adjustment; if they do, take it. Although inflation is only 2% per year, it adds up and it may not stay that low in the future.

There are different types of disability contracts: individual (private) guaranteed renewable, group disability coverage (if you practice with a group of physicians, or from your institution if you are salaried physician) or association disability coverage like BCMA or MD Management.

Before the insurance company issues you a policy, a detailed questionnaire about preexisting medical conditions or illnesses, medication, family history will need to be completed. Your information may be reviewed in more detail if the insurance company needs to asses risks. You will also be required to have a medical exam and a blood screening. The insurance company is issuing an amount of coverage based on your income and you may also need to provide a previous income tax report. With this information, the insurance company will determine the cost of your policy.

Depending on who paid for the benefits and whether they were deductible for business expenses will determine taxability. You may want to verify this question with your accountant before assuming that your disability benefits are tax-free.

Consider discontinuing your disability insurance when you reach 60 years of age, even more so if you consider a pre-retirement, or to work part-time in retirement. It sounds risky and like it could be inappropriate advice, and your agent may try to convince you out of this decision. But review your financial status, your obligations and your debt, and explore the possibility that you may save money by discontinuing your disability insurance policy.

As you are enrolled in a Royal College approved residency program, you have medical liability coverage. Residents in training should probably be buying disability insurance to protect their future earnings.

Life Insurance

You will often see recommendations that all adults should have life insurance for reasons like protecting their spouse, children, mortgage, in the event of a divorce, loans, business and estate taxes. There are certain circumstances where you may not need it. If you are not married and have no dependents, it is probably not required until you get married and have children. The policy may not be necessarily less expensive because you purchase it at 30 or 40 years old and if you have no risk factors. When you have reached pre-retirement age (60 years old) and you are financially secure, you may decide that life insurance is not required. Also, premiums rise dramatically at this age because of your increased health risks.

If you own a home, are married, have children, then you should have sufficient insurance to cover at least your debts. This could include the amount owing on your mortgage, car loans, student loans, credit cards and your own income tax. You should also have life insurance for your spouse if he/she is not working. If he/she is working, your spouse should have life insurance coverage, as there are always expenses, disruptions, children care and a certain amount of debt.

The amount of your spouse's life insurance should be based on their salary and their contributions to the household debt. A coverage amount for your life insurance policy of $1M should be reasonable.

Although the loss of a child is an unbearable thought, and can cause great pain, there is no reason to purchase life insurance on a child. Life insurance is not an investment for college. A life insurance policy purchased for your child will not decrease his/her life insurance cost or enable him/her to purchase more insurance when they become adult. This could be a waste of money. It is a better option instead to start a college fund for your children.

Life insurance is a contract between an insurance company and an individual to pay a specified amount of money to a named beneficiary upon the death of the insured. The amount of money is predetermined at the time of the application for the insurance. There are two principal types of life insurance policies: term life insurance and permanent life insurance.

Term Life Insurance provides specific benefits (giving high value and short term protection) to your spouse or beneficiaries. The premiums are lower and there is no cash value to it. Term plans may also be renewable after 5, 10 or 20 years without providing proof of health. The price increases at the same time as the age of renewal. You can purchase term life insurance as an initial step before purchasing permanent life insurance.

Term insurance can be purchased with level premiums from 1 to 30 years. After that time period has expired, the insurance will become very expensive. There will be a need at this time for reassessment of your risk factors and assessment of your insurance.

Permanent Life Insurance provides lifetime death benefits to your spouse or beneficiaries upon your passing. The premiums are higher. Because permanent life insurance has a cash value, it is a real asset and provides the possibility to borrow against the cash value. After two to three years, you will see a positive value in the cash value of the asset. You may also get some money back if you cancel your life insurance. You can also use the cash to pay for a premium on your permanent life insurance, should you choose to increase it. But your are paying the insurance company money/fees every year with uncertain long-term return, except if unfortunately you die and the insurance company pays your spouse/beneficiaries the full amount of the policy. However if you decide to cancel your $1M life insurance policy at 60 years of age because you are financially secure or you decide you do not need it anymore, the cash value may be less than you expect. When you initially purchase your insurance, you should verify the future cash value of your policy in 20 or 30 years. The concept of permanent insurance, as opposed to term insurance, is that for permanent insurance, you pay a level premium throughout the duration of the contract, but the level will stay the same throughout your life span.

There are many other types of insurance vehicles, such as Participating Life Insurance, Universal Life Insurance, Millennium Universal Life Insurance, Critical Illness Insurance, Covered Critical Illnesses, Life Style Protection Plan and Estate Tax Insurance Plan, just to name a few. You really need an informative session with your insurance advisor to understand what the company offers and which insurance plan will be beneficial for you and your family.

Extended Warranty

It seems that whatever we purchase these days, we are being asked or offered to buy an extended warranty. Any mechanical or technological product comes with an automatic manufacturer's warranty, covering some or all parts of the product for a period of time. It is a good idea to keep the warranty documents in a safe place, as they may be useful one day. But with your purchase, the salesperson will often recommend an extended warranty. Truly, in most cases, extended warranties are rarely useful. The company selling the extended warranty realizes that this can be an added profit over and above the profit from selling you the initial merchandise. You are better off buying a quality product from a reputable company and taking good care of it. This will save you money.

For example, I bought a new Apple IMac and was asked to buy an extended warranty of 3 years. First, Apple IMac computers are generally well built, and do not usually break down, secondly, the company comes up with new products frequently which will most likely make my IMac obsolete before the extended warranty is done.

The same logic applies when purchasing a car. Most car companies have 100,000 km or 3 years warranty covering most of the car parts. The way cars are built today and if you take good care of your car, the car should last more than 100,000 km with low maintenance costs. These costs will most likely be less than your extra-warranty annual cost.

Although washers, dryers, air conditioners, etc. are expensive, they also do not require an extended warranty. Appliances are built to provide good service for 15 to 20 years or maybe even longer. In summary, always buy quality, not just performance but durability, dependability and affordability.

Travel Insurance

Most people travel, personally and with family, for work, presentations/lectures, medical meetings, and for holidays. We have an extremely good and comprehensive health care system in Canada but out–of–Canada emergency medical expenses are generally not covered by our provincial medical plan and according to which country you travel to, these medical expenses can be quite high. The USA can have very costly medical expenses.

There may be travel insurance coverage provided on your credit cards, like car insurance, but it is very limited and you should confirm what type of coverage you have with any of your credit cards before assuming you have sufficient travel insurance.

As you get over 65 years old and you add some risk factors, medication and preexisting health issues, travel insurance starts to cost more. This does not mean you cannot be insured, but it may cost a premium, per day or per year according to the travel plan you need. A travel insurance package for the year may be less expensive than you might think. This will keep you and your family secure with travel coverage when travelling. You should not leave on your trip without some form of travel coverage. There are many insurance companies offering different travel insurance plans, but I would advise two companies, each for somewhat different reasons: Canadian Automobile Association (CAA) and Canadian Association of Blue Cross Plans.

The Canadian Automobile Association is a non-for-profit federation and was founded in 1913 *(http://www.caa.ca/).* It provides extended roadside assistance services, automotive and travel services, multiple insurance services and member discounts. It is affiliated with the American Automobile Association. Every Canadian province has an affiliation with the Canadian Automobile Association. The roadside assistance program is excellent and provides not only road assistance but also a network of inspected and approved auto repair facilities. It is worth having a CAA membership just for road assistance and member discounts (such as airport parking and hotel booking). The membership is $100 per year.

"I parked my car in the long term parking at our airport. I picked up my car after 10 days and the bill was $99.58 but with my CAA membership card, I paid only $56.38".

It takes only a few airport parking tickets to save the same cost as an annual membership. CAA also provides several travel insurance packages for individuals and family groups; you can choose from a package of multiple trips/entries and an annual plan for a very cost effective fee. It is recommended to that you meet with an agent to determine what your needs are and read the fine print on the policy. From my personal experience, there are several services they do not cover, like flying home with a sick family member and certain medical expenses like medical supplies. Their claim process is somewhat confusing and slow. Worth noting, CMA does provide a range of home insurance plans, extended health and dental insurance coverage, life insurance, help in providing the best car insurance coverage and travel/vacation packages (booking, trip planners and road maps).

Blue Cross member plans operate on a not-for-profit basis in Canada. All membership plans are associated with the Canadian

Association of Blue Cross Plans (CABCP) *(http://bluecross.ca/en/index.html),* which is linked with the Blue Cross and Blue Shield Association in USA and worldwide through the International Federation of Health Funds. The independent regional Canadian organizations offer a complete line of additional health, dental, travel and life insurance, aswell as disability income plans on an individual or group basis. Different travel insurance packages are offered on a per day, per trip, or annual basis. The forms are easy and clear to fill and customer service is impeccable. Their coverage is more extensive than CAA for a similar price and having dealt with Blue Cross for many years, there has been no problem in dealing with claims and obtaining reimbursement. Worth noting, is that if you have for instance, you had a cardiac catheterization and a stent implanted, Blue Cross will not insure you for your heart or heart related events for 6 months. They will require 6 months of health stability before providing you with full insurance that includes your heart or cardiovascular system, with obviously a premium that you will have to negotiate. Having said that, you can be insured for travel medical emergency expenses such as breaking a leg, catching the flu, having pneumonia or an operation as long as it does not relate to your cardiovascular system.

It is advisable to purchase the annual travel insurance for a nominal cost. Then you can relax when you travel and have a good time on your business trip or holiday with your family.

REFERENCES

Websites

1. http://www.icbc.com/
2. http://www.caa.ca/
3. http://www.bluecross.ca.en/index.html

PART 3:
Support Personnel and the Realities of Practice

CHAPTER 15:
A Travel Agent in a Wired World

Technology and Access

In the not so distant past, the Internet was a dream and travel was not as easy as it is today. Before the Internet the only help was available through travel agencies, holiday companies, and intermediaries to make travel reservations and other arrangements. The travel industry got a major boost with the deregulation and consequent boom of the airline industry, along with the development of the Internet.

Very rapidly, technology has allowed us to see the world. The world is at our fingertips with competitive and affordable flight tickets, hotel deals and booking and potential holiday experiences almost anywhere around the globe.

Cruise down the Yangtze river or the Danube river, bicycle in Burgundy, Corsica or Argentina, trek in New Zealand, Tibet or Peru, ski in Vale, Colorado, or Verbier, France. You can fish in Northern British Columbia or off the coast of Florida; try wild life watching in Galapagos or Africa. All of these experiences (and many more) are available at a reasonable cost. In recent years, travelling with children has also changed. Now, it is more affordable, there is easier flight access, computer amenities (games, movies, readings), and child friendly destinations while safety has also improved family experiences. Distances have been reduced with long haul flights to Asia, Europe, Africa, Middle East, and Australia.

The Internet can provide amazing choices, availability, and easy access, but it is easy to feel overwhelmed by the massive amounts of information and all the choices. The number of websites is endless; see below for some of the best.

LIST OF TOP INTERNET BOOKING SITES:

Kayak.com
Yapta.com
Priceline.com
TripAdvisor.com

Hotels.com
Expedia.com
Travelocity.com
Orbitz.com

Role of a Travel Agent

Is there still a role for the travel agent? Travelling for work and travelling with your family for holidays are somewhat different. The planning and organization, booking flights and hotels and the time involved doing these tasks can be quite time consuming. You can save valuable time by using a travel agent for both pleasure and business trips.

Work related travel is usually less complicated because there is less research involved in the process and it is more straightforward. For instance, imagine that you are attending an international conference in a large North American or European city, in a large well-appointed hotel. The dates are set and the conference registration package will mostly include the conference registration and the hotel registration. You will need a direct flight. Flying the day before and returning the day after the conference. On another occasion, you may receive an invitation to be a guest speaker at a national or international meeting, with all the travel arrangements made for you by the host. And along the way, there will be a multitude of small trips for a number of professional reasons as your practice evolves, develops, grows and you take on more tasks. They may be administrative, educational, you may be a part of a national and/or international committee and perhaps even fundraising.

All these commitments will require travel arrangements and with the progressive changes in the travel industry and the development of the Internet, there are 4 common approaches to making travel arrangements:

1) Make all the bookings, yourself

2) If your spouse has time, they can make the travel arrangements for you
3) Your secretary can make the travel arrangements
4) You use a travel agent.

Interestingly, these four common approaches can apply not only to your professional travels, but also for your family holidays.

The Pros and The Cons

Here are the pros and cons for all of these approaches:

As a professional, you work in excess of 50 hours a week, and often more. Depending on your practice, you may be on-call at least once a week. Over and above your patient load, you will likely have administrative duties, and possibly educational and research projects. You will also have a personal life to balance, probably a spouse and children. In addition to all of these commitments, try to find a little time for yourself. Juggling the time to book your professional travel and organizing a two to three week holiday for your family is time consuming, because it involves several emails, adjustments in the bookings and this can ultimately lead to more stress for you.

" Jim was flying to Hawaii with his wife and children. He wanted to book all the tickets on his points. The family was excited at the prospect of two weeks off, they arrived at the airport only to realize they were there on the wrong date. The flight was the next day and that they had actually arrived at the airport 32 hours before their flight was set to leave. Jim was so busy that he got the dates wrong when he booked the tickets."

You can decide if your busy schedule leaves enough time for the complexity of making travel arrangements. It might be better and easier for you to have a travel agent make all the travel arrangements for you.

Your spouse does not work, if you have young children, or he/she works part-time and his/her schedule allows him/her to organize your professional travels and family holiday travels. It is a good solution for you and will save time. Your spouse can make all the arrangements, look for the best deals and make any last minute changes before your departure. It works for many couples and travel plans can be done easily.

"Mary had organized a two week family holiday to Barcelona and the south of Spain. Arriving at the hotel in Barcelona, they realized their hotel was just out of town, although they thought they were in the center of Barcelona. Mary had not realized that there were two hotels with a very similar name. Still, they enjoyed the countryside."

It does not necessarily mean that a travel agent would not have made the same mistake or recognized there were two hotels with a similar name, but as they do this work all of the time, they may have been able to avoid these types of errors.

Issues to Consider

Given the information that is now available on hotel booking sites on the internet, complete with Google maps showing the location of the hotel in relation to the city you are travelling to, research before you leave to avoid the same mistake!

Your secretary can be very helpful in booking your professional trips but not all secretaries will do this for you. Many unionized or university appointed, or hospital appointed secretaries will not do trip bookings, as it is not part of their job description. Your private secretary may be more inclined to organize your professional trips as it is part of managing your office, as well as paying bills, organizing visiting professors and other administrative duties. This approach can be very useful in managing your clinical time. My secretary did not organize any of my travels, but did so for my associate.

A travel agent does offer multiple options/possibilities. The agents provide one-stop shopping for travel related products for flights, hotels, car rentals, cruises, tour packages, etc. They have access to a reservation system that quickly and easily shows multiple options that we do not have access to, despite the multitude of Internet websites. With frequent flight cancellations, surcharges and other things going on, a travel agent may be very helpful to navigate the travel landscape. Airlines are offering less and less seats to redeem your frequent flier miles and are adding more restrictions that an agent can help you with. An agent can also help you with entry visas wherever necessary, so you don't have a problem at the immigration area. You must remember to check on the website of the country you will visit if you need an entry visa or medical shots to visit the country.

Travel agents provide recommendations based on their knowledge of a client's preferences and lifestyle. I have had the same agent for 25 years and she knows my preferences for my flight seat assignments, car rentals and the type of hotels I prefer. Travel agencies sometimes have access to special deals or a bonus that may not be available to the general public. On my last trip to Australia, my travel agents were able to get a better price for me on my hotel than what was quoted on the Internet. If a problem occurs on any trip, the travel agent knows who to contact and will do their best to help you resolve the issue with haste.

"I was in Shanghai as an invited professor and I had several lectures to give. During the conference, I received an email message about a family emergency. I sent an email to my travel agent and briefly described that I needed to fly back home the next day. I finished my last lecture and in the evening, I had the confirmation of my flight change."

For long holiday trips, both my wife's family and my family knows the contact details of our travel agent in the case of an emergency. This is an easier option, for the travel agent to reach us, if we are out of the country. Our travel agent knows the timing of the trip, schedule and hotel reservations.

Fees

The price is not always the bottom line. Most airlines and other travel industry partners have either eliminated or drastically reduced the compensation they used to provide to retail travel agencies, so travel agents do charge a fee for their services. The fee is dependent on what is being booked, whether the airline/hotel/cruise line compensates the agencies and the time involved in researching options and completing the booking process.

As an example (and it does not mean it is the industry fee standard), my agent's fees start at $30 + GST for a relatively simple domestic booking. Fees for international bookings range from $60 + GST to $150. + GST, dependent on whether there is any compensation from the airline, or other suppliers the agent is using, and the time involved in researching the trip options. It is up to your travel agent to determine with you what the fees will be. The travel agent will let you know the fees involved after discussion about the trip, bookings you want to do and research for the actual trip.

In my case, I believe a fee of $150 + GST to organize a 4 to 6 week holiday in another country, book international flights, reserve several hotels and possibly a car rental, is a small fee on a trip costing several thousand dollars. It is my personal opinion, but I believe it does make sense to work with a travel agent you are confident with, when you have a busy practice and a busy family life. You probably do not have the time to spend hours on the Internet organizing your work travel or plan a family holiday. The use of a travel agent will allow you to spend precious time with your patients and/or family.

I believe that a physician's time is valuable and surrounding yourself with excellent, reliable and professional assistance provides multiple advantages. A travel agent can help with all travel issues. For instance, I limit my time on the Internet. I only provide my travel agent with my hotels of choice and price range. For example, I send my agent an email with my detailed itinerary, travel dates, airline to compare fees from and the number of days per destination. When all of the travel arrangements have been confirmed by email, (including visas if necessary), my agent sends me the travel package as an email attachment. This is my approach to travel; I do find that it makes the process of planning and organizing travel that much easier.

Finally, I do not believe you will have the time to interview several travel agents, as you try to find a suitable travel agent for all your needs. Probably the best way to find a good travel agent is to talk to your friends, friends of friends, family, and colleagues about their experiences with their own travel agents. You may need to work with several agents before developing a trustworthy relationship with the right one. But you also need to remember that an agent will be as good and efficient as the information you provide. Happy travels.

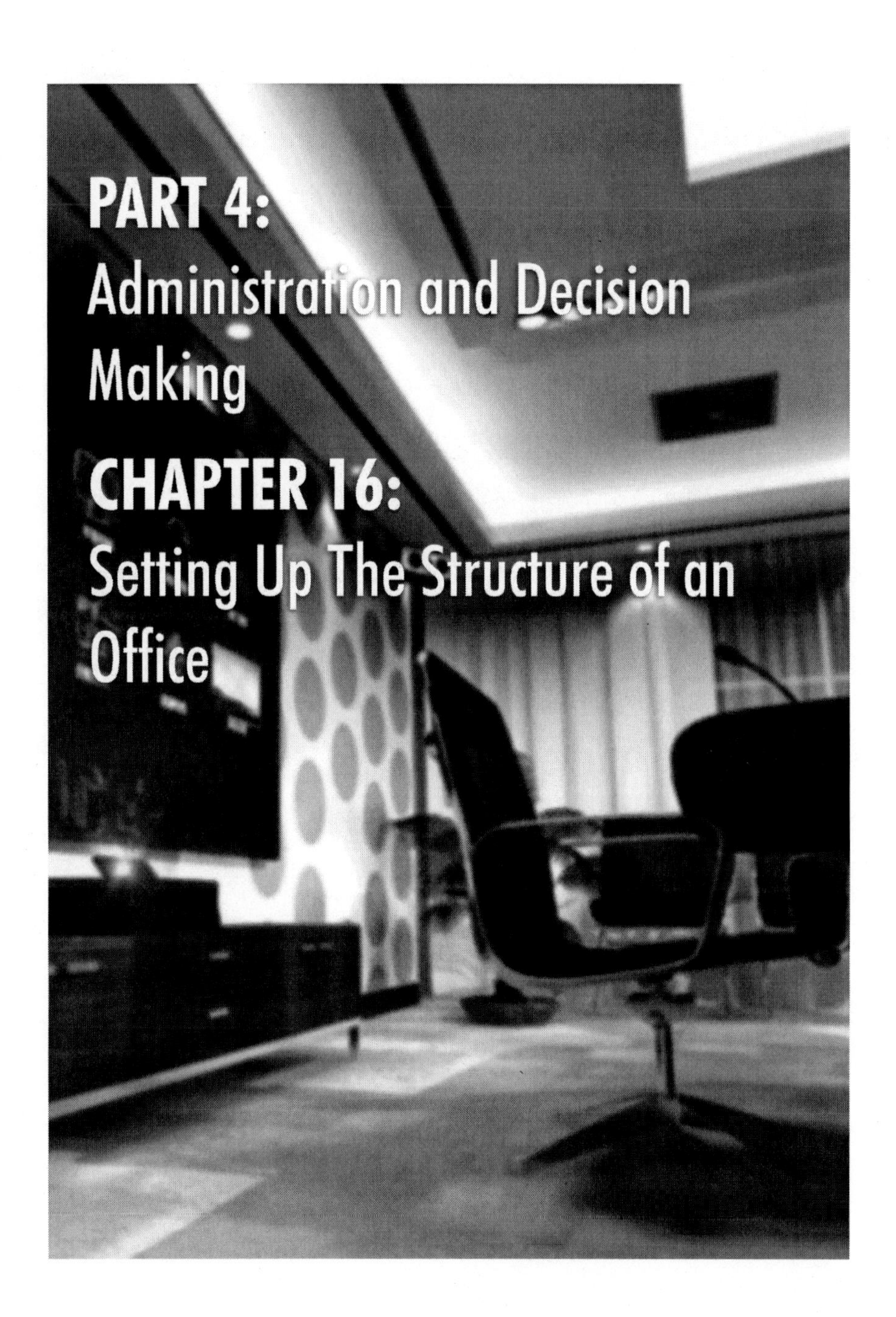

PART 4: Administration and Decision Making

CHAPTER 16: Setting Up The Structure of an Office

Setting up your office is an important step. An efficient and comfortable working environment facilitates efficient service delivery to patients, less stress for you and your staff, professional satisfaction and increased income.

Physicians' office requirements can vary greatly depending on the doctor's specialty. A community-based family physician, paediatrician or internist will have much more complex office requirements than an anaesthetist who works in an academic hospital.

Decision-Making

Regardless of your specialty, whether you work alone or as part of a group, if you are self-employed or an employee, you have a vested interest to ensure that your "home away from home" is professionally comfortable, managed effectively and efficiently. If you join an established practice or work as a salaried employee in an institutional setting where an office is provided for you, you can negotiate for an office that will meet all of your personal and professional needs.

Most of the doctors who go into practice after residency training will be in a family practice, medical or surgical subspecialty practice. You may be hired to join another physician, a group of physicians, a hospital-based practice or a purely academic practice. For any of these situations, you will not need to set-up an office from scratch. However, you may need to simply add your own personal/equipment/computer technology and furniture to the space.

The benefit of working within an established practice, group practice or hospital-based practice, is that you do not need to create a full office as many ancillary support systems are already in place. A group practice also reduces the costs per physician of both supplies and capital equipment. By consuming larger quantities of supplies, the group can generally obtain better pricing. Costs can also be reduced through sharing fixed-cost expenses such as: office space, automation technology, medical equipment and communication systems. The investment may be less and you can start to work sooner.

Logistics

This chapter will be more for physicians starting their own office. Some information may be useful to everyone. The chapter is only meant to be a guideline for what you may need to establish your office. It will not provide all the details, as this information can be found in many available books, support programs at your local provincial medical association and Colleges.

There are many things you need to think about when establishing your own medical practice and each practice is different. Part of that decision will be based on where you want to live and whether or not you want to commute. Do you want to live in a big town or a small town? Do you want to be affiliated with a teaching hospital? Please refer to Chapter 6, as this section of the book will address these considerations.

In order to help you realize your goals, it can be extremely beneficial to assemble a good support team of professionals including: an accountant, a banker, a financial adviser, a lawyer and a real estate agent.

Once you've identified your preferred practice type and the community you'd like to work in, then there are business considerations to think about, such as choosing a specific practice location.

After you've selected your ideal location, the next step is to negotiate your lease. You'll need to consider several issues including: net rental costs, taxes, maintenance and insurance costs, the lease term, the landlord's responsibilities, your responsibilities, liability issues and hours of operation, etc. Essentially you are creating a business plan for your practice.

Assuming a practice may require you to buy existing equipment from the departing doctor, you may also incur additional start-up fees (If the capital expenditures of the outgoing physician's associates or partners are not fully depreciated. These costs are usually not significant and can easily be financed).

What are the benefits and risks of assuming an established practice?

1. The practice may have comprehensive medical records of patients who have realistic expectations of the services you can or cannot offer
2. An effective, efficiently-run business that is professionally and financially rewarding
3. Experienced staff who already know the patients
4. Consider assuming a practice if you have already given it a test-drive through a locum risks
5. It may be a poorly run practice with patients who have unrealistic expectations
6. You could inherit the outgoing physician's potential mistakes

When you have made a decision and found a location for your practice, you will then need to plan the layout of your space, this can be a difficult exercise.

If you are new in a city, it is a good idea to consult with a commercial real estate agent to find a suitable space or a space that has been previously used as a medical office. This way the renovations can be kept to a minimum. The real estate agent may also be able to connect you with a qualified contractor for the renovations. Using a reference from trusted friends or colleagues is always the best option, however, if you don't have this option, then speak to the contractor's previous clients about their experience. You may also want to review some of the completed projects done by the contractor. This can be an effective way to evaluate a contractor's work.

The Physical Environment

Most new physicians will join a group practice in an office clinic setting, rather than design and build a medical office from scratch. You should base the evaluation process of your medical office in a similar way that you would as when you are considering your own home. This evaluation process could include features like: comfort, functionality, personality, accessibility and street appeal.

Learn how to evaluate your work environment during your residency. Assess the current offices and clinics you are working in. Establish your "ideal office." Note the spaces, the configuration, the design, the equipment, the office space for staff and doctors. Keep remarks about what you do like and don't like in an office setting. If you are joining an existing practice this exercise is still worth the effort. It will give you an idea of the environment you will feel comfortable working in.

When evaluating an existing practice or designing an office from scratch, you may want to consider the following:

- The Public Access Area
- The Waiting Room
- The Reception Area
- The Common Area
- The Examination and Procedure Room Area
- The Physicians and Staff Offices Area
- The Office Equipment
- The Supplies Room
- The Telephone /Internet System
- The Electronic Medical Record System
- Infection Control

The Public Access Area:

The public access area refers to the space that the public has access to, in and around your office location such as: the parking area, wheelchair access, elevators, corridor size, diagnostic and other allied health services and the emergency 911 team. It may not be high on the list, but it may be useful for yourself, your team and your patients to have a coffee shop or small deli near the office.

The Waiting Room:

The waiting room should be aerated and brightly lit so the patients can read and complete forms that may be required. It must also be spacious enough to allow for the number of patients the practice has on any given day, plus a magazine rack or patient information display.

Having furniture that is comfortable, practical, and ergonomically correct is beneficial for patients and staff. Arranging the furniture in a manner that promotes ease for patients and has a pleasing effect is beneficial. Most furniture can be bought in a very affordable way from stores such as IKEA, Staples and Office Furniture. In addition, many offices now provide TV and/or Internet and music. Finding a colour palette for the entire office makes choosing colours and fabrics much easier and will produce a better effect. Enlisting the support of a trained interior designer may be useful, save time and save costs. This could result in a finished product that is pleasing for your patients and staff, as well as durable and low maintenance.

The Reception Area:

You want to aspire to have an efficient, effective and ergonomically designed reception, administrative and clerical space in your office. This will significantly benefit your staff, your patients and yourself. During residency, most physicians have minimal exposure to office operations outside of the examination rooms. To create the best office space, it is a good idea to incorporate the feedback from your nursing, clerical and administrative staff.

The Examination and the Procedure Room:

During your residency, you probably experienced work in a dreary, small examination room with inadequate lighting and outdated equipment. This created an uncomfortable environment for both you and the patients. So, during training it would be wise to make detailed notes about the procedural rooms you liked the most.

The reality is that over the next several years you will probably spend more time in your examination rooms than you will in your home kitchen or family room—so invest in your practice environment and make it functional and comfortable. Consider some of the following ideas when you organize your examination or procedure rooms.

The examination room is actually the interview room, the examination room and in some cases the procedure room as well. Because the space will be used for different purposes, it should be spacious enough to accommodate chairs, an examination table, a workstation for chart completion, a computer, sinks, equipment, supplies and the movement of people. A room that feels crowded can be an uncomfortable room for both you and the patient.

If all of your exam rooms function as multi-purpose rooms, it will make your workspace more efficient. Then neither you nor your patient will need to wait for a particular room to be available. The rooms should ensure that the patient can disrobe in privacy and you should offer your patients gowns that are adequate to maintain modesty and warmth. Also, a place for the patients to hang their clothes is important in the space. Position the examination and procedure tables (and chairs) so that both the patient and physician can access them comfortably. Most physicians are taught to examine the patient from their right side and as a result position the examination table to the left side, parallel to the wall. If you do procedures that require access to both sides of the patients, consider a larger room and having the examination table available so it can be set in a locked in position. It is a good idea to ensure that the room can accommodate computer upgrades for electronic medical records (if it has not been upgraded already). Finally, if you need further assistance, national and provincial specialty associations often have resources available to assist with the design and furnishing of examination and procedure rooms.

The Physicians Offices:

A physician's office does not need to be large. It is customary that you will meet your patient in the examining room. Space is required for a desk and comfortable chair. There may also be additional items that you will need space for such as a computer area, filing cabinet, bookshelves and a spot to hang your clothes. As previously mentioned, for the reception area and common space, you should choose a soothing wall colour for your office. A window and natural light can also be a great asset to any office. These are only suggestions, and of course you can make adjustments depending on your personal needs and taste.

Office Equipment

The Supply Room:

It is beyond the scope of this chapter to provide a detailed inventory of specialty-specific medical equipment, office and medical supplies. There are several websites that will provide this information for you. Additionally, the references at the conclusion of this chapter will provide you with details of where you can purchase supplies and equipment. There are many more options that you may find through colleagues, your own provincial College and the Internet. You are encouraged to ask the physicians and managers of the clinics where you work as a resident if they can share with you more information about the organization, equipment, supplies, suppliers, etc. Provincial and national specialty associations often have resources that can help new physicians organize their offices too.

Ultimately, a well organized office, spacious waiting and reception areas for patients, properly equipped examination rooms, staff that are polite and professional, will make your practice efficient and enjoyable. Be thoughtful and aware of nuances in your early years, so you are equipped if given the chance to make your own space later.

REFERENCES

Canadian Medical Supplies Companies:

1. canmedhealthcare.com
2. medicalmart.com
3. stevens.ca
4. stathealthcare.com

Books:

1. Tom Harbin
   The Business Side of Medicine: What Medical Schools Don't Teach You?
   Mill City Press Inc., Minneapolis, 2013
2. Judge Huss, Marlene Coleman
   Start your Own Medical Practice
   Sphinx Publishing, Naperville, Illinois, 2006
3. C. James Holliman
   Resident's Guide to Starting in Medical Practice
   Williams & Wilkins, Baltimore, 1995
4. C. Rainer
   Practice Management: a practical guide to starting and running a medical office
   Wyndham Hall Press, 2004
5. J. P. Daigrepont
   Starting a Medical Practice
   AMA, 2th Edition, 2003
6. Infection Prevention and Control for Clinical Office Practice
   Provincial Infectious Disease Advisory Committee (PIDAC), June 2013

Websites:

1. https://www.cma.ca/En/Pages/new-in-practice-guide-2014.aspx
2. zmags.ca/2014/new-in-practice
3. zmags.ca/2013/new-in-practice
4. physicianpracticesetup.com

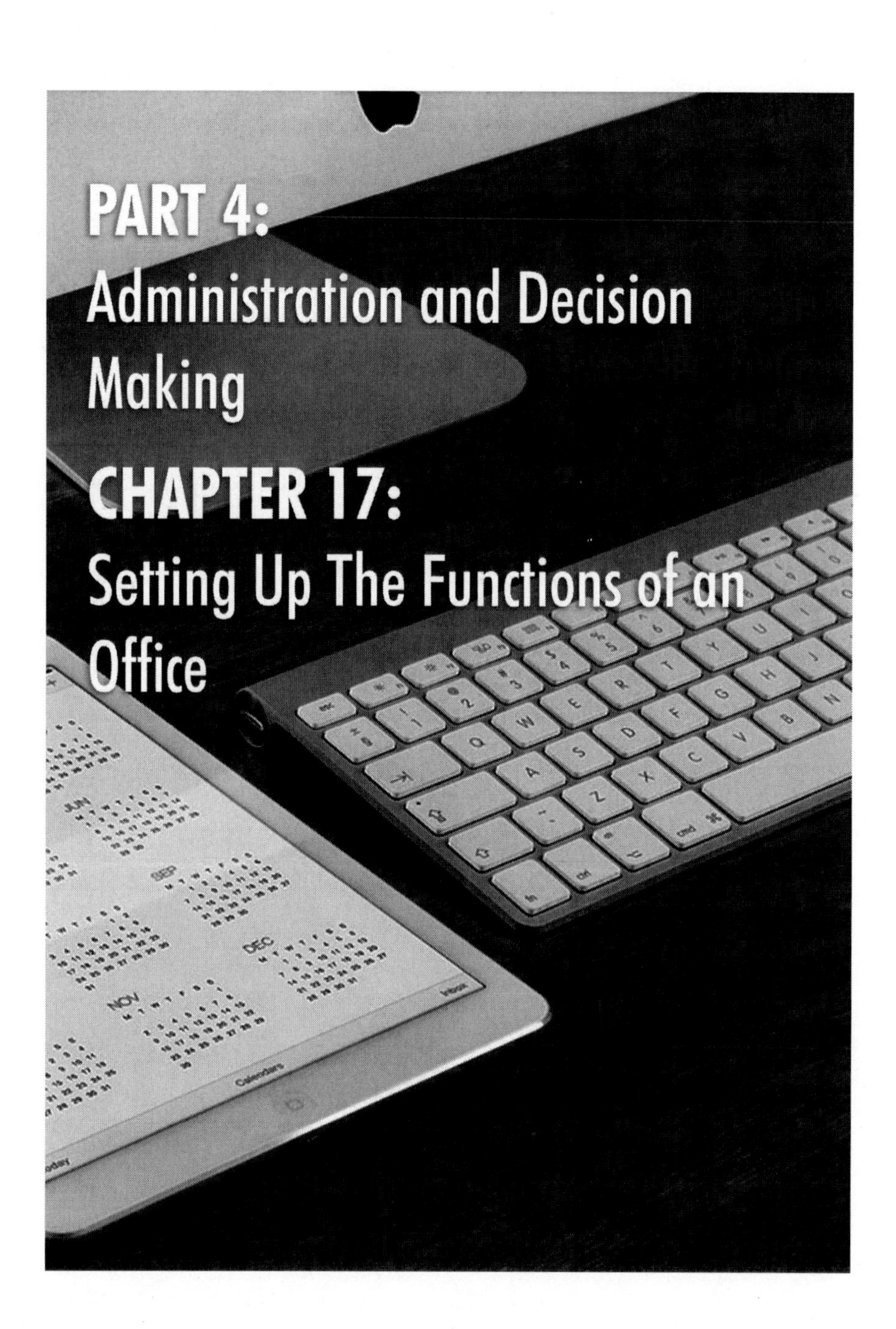

PART 4: Administration and Decision Making

CHAPTER 17: Setting Up The Functions of an Office

Fax, Copy Machine, Scanner

Fax machines can be used for routine tasks such as: requesting consultations, booking diagnostic procedures, transmitting or receiving test results and communicating with pharmacies. Fax machines are a cost-effective and time-efficient addition to your medical office communications.

Whenever possible, your office should use faxes instead of telephone calls. Not only will you have a written record of the communication, both you and the recipient can deal with the fax when it is convenient. Because the fax machine will be used frequently in your office, it should have a dedicated fax phone line. Essential functions of a fax machine include: speed dial, automatic redial, delayed transmission programming and auto-sizing.

Similarly, the photocopier has become an essential piece of office equipment. A common use of the fax machine is to use it for copying and transferring medical records. For maximum convenience, the fax machine and photocopier should be located in the reception area. Scanners are essential for offices that utilize electronic medical records. Many diagnostic and laboratory centres, pharmacies, hospitals and physicians' offices do not have the capability to send all of their communications electronically and these reports will need to be converted into electronic form for an EMR system. One piece of office equipment that combines fax, photocopying and scanning technology will prove to be the most practical for your office. They can also be purchased quite cost effectively.

The Telephone and the Internet System

Your telephone system is the lifeline of your medical practice. Your phone and Internet system will be quite complex. You will need to manage incoming calls from patients, dedicated lines for outgoing calls, a private office phone, an internet line and the fax-modem line (this will be used to send bills to the Ministry of Health by electronic data transfer). You will need to ensure that the system has the capacity for additional lines if your practice grows. There are two schools of thought about having telephones in the examination rooms. Some physicians feel that they should not be interrupted while with a patient and that office confidentiality could be breached should staff or patients use these phones. It is also important and easy, to incorporate functions that prevent patients from making long-distance calls or listening to other conversations.

The other perspective is that it can be effective and efficient to have access to a phone in all work areas. For instance, reception and administrative staff can help patients to rooms without being far from the phone—which will allow them to take calls that otherwise might go unanswered or would go to voicemail. Sometimes physicians need to interrupt a patient consultation to take a telephone call, but it is easy to standardize the protocols so that staff knows when it is appropriate to interrupt the doctor and how to maintain confidentiality.

Taking the call when it comes in saves the time-consuming task of having to call patients back. What is probably more important than a phone line in the examination room, is a computer terminal and keyboard for you to access patient information, test results and enter your notes after seeing a patient. Access the resources that cma.ca offers.

The Electronic Medical Records (EMR)

Medical physicians are a self-regulated profession in Canada and they are held to certain professional standards. These standards, enforced by the various provincial physician college licensing authorities, apply not only to medical competency, but also to keeping medical records. Physicians are the trusted stewards of their patients' medical data and they have obligations to ensure that this data remains private, protected, intact, well organized and available when you need it. It is the physician who is responsible for the integrity and security of their EMR system.

You know about the benefits of using an EMR and you have probably also heard about the difficulty about choosing and implementing an effective system. The future of the medical office lies in a chartless office and will increasingly transition to everything being kept in electronic medical records. Major improvements have been made to EMR systems and today more than 30% of Canadian physicians use some version of an EMR system. This percentage will increase as more records software systems become available. Implementation is becoming more affordable and in some provinces, financial assistance is available to physicians who are willing to convert their paper charts to a computerized system.

Unfortunately, the degree of provincial government support and resources for EMR systems and electronic health service delivery varies widely across the country. A further complication is the lack of standardization and common formatting that currently exists between the provinces. For example, a system that meets all of the Ontario requirements may require significant customization for Alberta. Accordingly, few EMR providers offer comprehensive national service; many suppliers are still small, regional companies. It has also been difficult to convince physicians who are currently in practice to invest in EMR systems. They are understandably hesitant to make the significant investment of time and money for an EMR system. They are unsure if this new system will meet their future needs.

But consider this, there are many frustrations with having a paper-based office. Physicians struggle with missing and misfiled charts, lab results, near-illegible handwriting, large and unwieldy lab reports, lack of storage and the sheer inconvenience of updating patient charts at the end of the day.

The advantages of an EMR system are many such as:

- You can enter your notes at the same time that you are seeing a patient and decrease the chart notes workload at the end of the day.
- Fewer calls from the pharmacists.

- You can treat chronic care or complex patients with more ease because you can easily track which tests were done, review lab work and trend them over time.
- You can use educational tools like graphs, charts and diagrams to illustrate to your patients the actual effects of the treatment.
- There are templates to make quick printouts during patient physical exams.
- You can access information from all members of the health care team (such as nurses, dietitians, etc.).
- You can provide preventive care with electronic reminders. An example of this is that you can provide group care by quickly sending a letter to all your patients over age 65 reminding them to come in for their annual seasonal flu shot.
- You will have a valuable database that can be referenced to improve quality of care in the future.

EMR Selection Tips

MD Physician Services has the following tips for physicians selecting an EMR:

- Use your time in locums to sample various products and learn the business side of EMR systems, including; billing, reconciling and preventive care features.
- Choose an EMR program that works the way your brain works – it should feel intuitive. Is all of the information that you need available on one screen, or will you need to click through several screens to find what you're looking for?
- Try to shadow a physician who is using EMR and observe.
- When you attend conferences or trade shows, spend some time with vendors at their booths. Try out their demo products and ask questions about the features.

- Don't learn simply how to use the record keeping aspect of an EMR – take the time to learn its messaging, billing and referral systems too.
- Ask a lot of questions – not just of the physicians who work in the practice but also ask the receptionists and nurses who use the tool on their jobs every day.

Common Features of Medical Software:

Electronic medical record (EMR) or electronic health record (EHR) software can help you create and store digital patient records, track patient notes, demographics, histories and medications. The features include e-prescribing, entering daily notes, E&M coding and much more. EMRs may also provide lab integration, device integration, tablet support and voice recognition. The patient scheduling feature automates the process of scheduling the patient's visit. The medical billing feature manages the creation of patient statements and submission of claims. Functions include; coding, electronic claim submission, payment tracking and reporting.

There arc resources available through different agencies in Canada that can help you sort through all the information that is available on the EMR. These resources include:

1. The Canadian Medical Association (CMA) *(http://www.cma.ca/En/Pages/cma_default.aspx)* provide resources, tools and they even offer a course that can assist you with the implementation of electronic medical records. They can also assist with some helpful privacy resources.
2. Canada Health Infoway *(http://www.infoway-inforoute.ca/)* provides financial and strategic assistance to help build electronic health record systems across the country.

There are many Electronic Medical Software programs available. For more information you can consult:

- Oscarcanada.org
- Capterra.com
- Cma.ca

OSCAR is an all-inclusive EMR software program, designed by doctors for doctors, for use in medical offices *(http://www.oscar.org)*. OSCAR is also used by a variety of health care professionals. OSCAR software can be downloaded for free by anyone and the source code is distributed with the software. This allows you to enable peer review and collaboration. It should also be noted that OSCAR is now the EMR used by Departments of Family Practice at a number of medical schools, including; McGill University, Queens University, UBC and McMaster University.

Overall, the largest numbers of practitioners use Oscar in Canada (Ontario, Quebec and BC). The true number of Canadian users is probably higher than estimated, because physicians can self-install OSCAR. Physicians can also be eligible for government funding in some provinces.

Capterra will provide a list of the most common EMR software programs available and include a directory of the pros and cons of each software program.

The Billing System

In a fee-for-service system, the physician is a self-employed professional who bills for each individual service provided according to the Fee Schedule Billing Guide provided by the Government. The main department responsible for payment for insured services is the provincial Ministries of Health. Other agencies may be involved in fee-for-service remittance like the Workers' Compensation Boards and some federal government departments (such as the Veterans

Affairs of Canada, National Defense, Indian and Northern Affairs Canada and Public Safety Canada). From time to time, other services may be paid by a third party such as an insurance company.

As recently as 10 years ago, the vast majority of physicians were remunerated solely by billing for services provided, but the concept of alternative payment plans appeared in academic and institutional settings, as it became obvious that monies generated from the provision of clinical services were no longer enough to remunerate the faculty for the teaching and academic administrative services provided. Academic institutions now negotiate for a "global budget" to pay for the faculties' provision of all services.

The variety of alternative payment plans now offered to primary care doctors in some provinces has encouraged many to move away from the traditional Fee for Services (FFS) payment model for some or all of their practice services. Understanding FFS billing is essential for all physicians, regardless of their payment model, for several reasons:

a) For the majority of family physicians, remuneration will still directly or indirectly depend on FFS billing.
b) Most Alternate Payment Plans (APPs) require shadow FFS billing for all services provided, so that the Provincial Ministry of Health (MoH) can track if there is a change, an improvement or a drop in services offered under the new payment formula. Shadow billing requires the physician to submit an invoice for all services provided as if still paid by FFS. This applies in Ontario, for example, where some family doctors work in such settings as family health networks and family health organizations. Shadow billing is also required of many specialists who work under APPs. To encourage compliance, bonuses for effective shadow billing are now being offered by some provinces. Most APPs require academic institutions to capture and submit shadow FFS billing for all services provided by faculty and residents.

Procedural fees are the bread and butter of specialists such as obstetricians, surgeons and ophthalmologists, and are especially important for anesthesiologists and radiologists, whose billing is mostly procedure based. However, forgetting to bill for minor procedures such as urinalysis, injections, phone supervision of anticoagulation and chemical treatment of skin lesions is very common among family physicians and this can result in the loss of thousands of dollars of income every year. Do not overbill but familiarize yourself with all the intricacies of fee-for-service billing.

Sessional fees are usually based on an hourly rate and pay for the delivery of specific services. For example, many emergency departments now offer physicians a guaranteed sessional fee per hour for working as the doctor on duty, regardless of the number of patients the doctor sees. The physicians may be obliged to shadow bill, so the actual services rendered can be monitored.

Bonuses are an increasingly important component for remunerating physicians for providing designated and targeted services such as complex care and evidence based management of chronic diseases such as diabetes, CHF and COPD. Financial incentives are not the same in all provinces, so you will need to verify your specific situation for your province. These bonuses can apply in different formats.

The bottom line is to approach all the services you provide with the checklist of what was the diagnostic code, service code and possible procedural codes and other fees and bonuses. By doing so, you will minimize the number of services you provided but failed to bill.

Finally, stay up-to-date with your fee schedule. The services that family physicians and specialists offer is constantly evolving, so new service and procedural fees are created to pay for these updated services. Failure to keep up-to-date with your fee schedule will result in a significant loss of income over your career.

As well, alternate payment plans are constantly changing and vary from province to province. Up-to-date information can be obtained at your provincial-specific resident association and medical association websites. Many provincial governments are now funding recruitment agencies that help residents, new entrant doctors and doctors in practice in not only finding work but also in understanding the varied payment models. CAIR — Canadian Association of Interns and Residents — lists all of this information on their website.

Infection Control for Clinical Office Practice

As a resident in training, we always worked in a hospital or office environment with an infection control policy in place. In many ways, it was such a part of the environment, that we may have not even been acutely aware of the necessity of implementing some infection control policy for a private office. As physicians, we need to protect not only our patients but also our clinical office staff and visitors as well. From both a structural and functional point of view there are ample opportunities for infection to be transmitted in a medical office setting. Infectious agents are not only spread person-to-person, but can also be spread indirectly through objects such as equipment. The waiting room of a clinical office can be a source for many communicable diseases. As such, protective mechanisms must be in place, not only in direct patient management but in handling of the clinical office environment as well.

The Best Practice Documents to prevent infection and control for a clinical office practice was developed by the Provincial Infection Diseases Advisory Committee on Infection Prevention and Control in collaboration with the College. This document is readily available and can be referred to, to prevent spread infections in your office.

These best practices document will outline:

- Principles of infection prevention and control in a clinical office setting
- Legislation relating to clinical office practice and duties of physicians as employers and supervisors
- Issues to consider when setting up a new clinical office

- Rationale and tools for screening and risk assessment for infection
- Recommendations for providing a clean clinical office environment
- Guidance for reprocessing of reusable medical equipment
- Protection and safety issues related to staff.

These infection prevention and control best practices for physicians are not intended to replace a physician's best clinical judgment, but will assist doctors with their clinical office-based practice. Some components have been derived from legislation and regulations, and will state in explicit terms what physicians should or should not do. Other sections are evidence-based best practices, intended to increase awareness about the day-to-day risks of infection acquisition and transmission in a physician's clinical office and to equip physicians with a practical guide and tools to minimize these risks.

By employing these best practices as part of routine care and knowing how to respond to the threat of infection in an expected fashion (e.g., implementing seasonal screening for acute respiratory infections), the risks associated with serious infectious disease outbreaks will be mitigated and the public will be protected by minimizing the risk of infection transmission.

Risk of Embezzlement

Embezzlement is not something that physicians may think about but it does happen. The best way to avoid this is to maintain some involvement with the record keeping and actively monitor the billing and flow of finances in your office (if your secretary or staff performs all of the accounting). It is easy when you are busy to ignore the paperwork and the finances. But do not neglect this part of your office. You should be sporadically checking the billing and cash flow of your office. Every six months or at the end of your year end, your accountant should reconcile all of the paid expenses and received income. This will ensure that you can effectively monitor the cash flow in your office and have a true understanding of your finances.

REFERENCES

1. J. Tse et al.,
   How Accurate is the Electronic Health Record? A pilot study evaluating information accuracy in a primary care setting Studies in Health Technology and Informatics, 168: 158-64, 2011

Physicians Electronic Medical Record:

1. http://www.oscarcanada.org
2. http:/www.capterra.com
3. http://www.cma.ca/En/Pages/cma_default.aspx
4. http://www.infoway-inforoute.ca

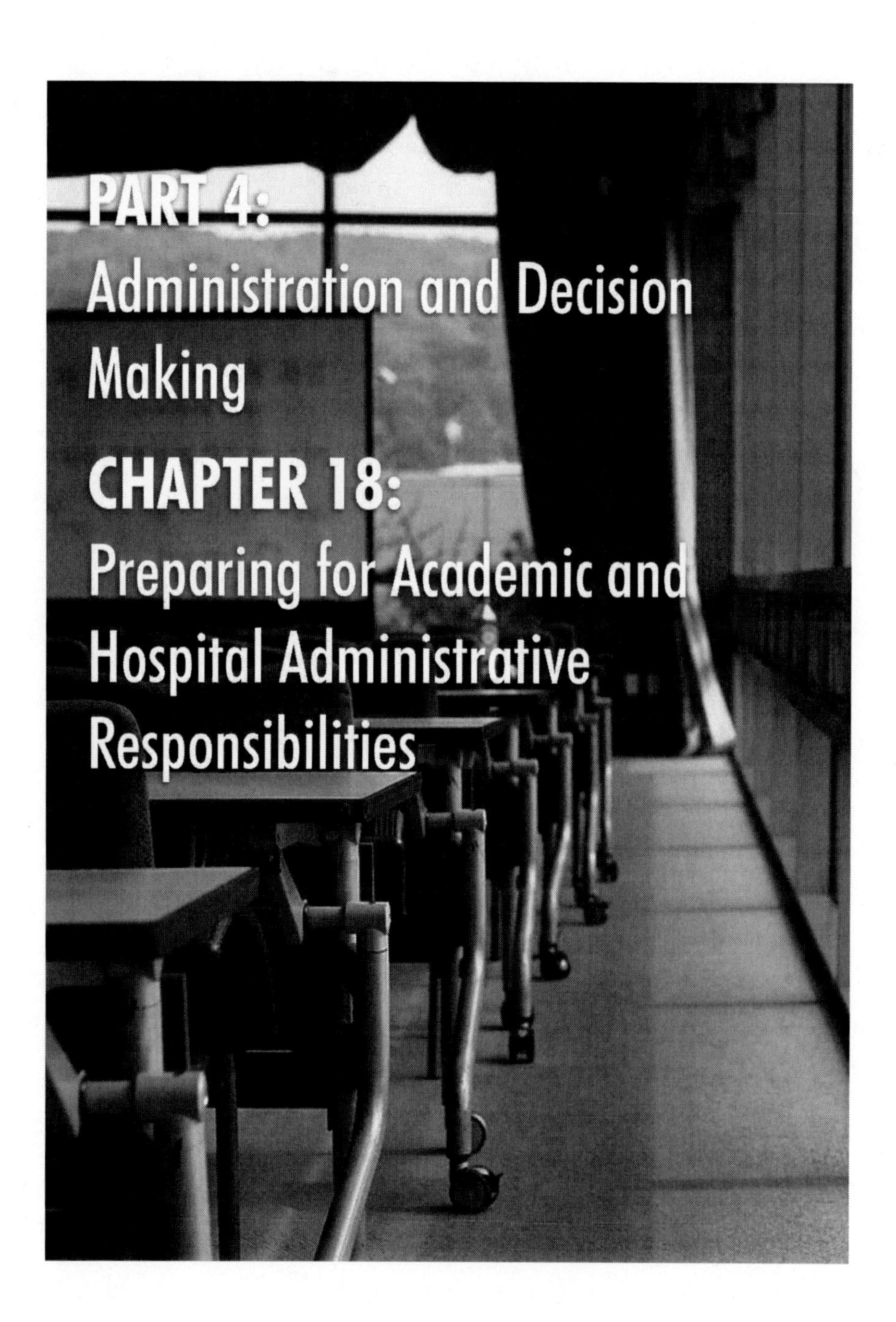

PART 4: Administration and Decision Making

CHAPTER 18: Preparing for Academic and Hospital Administrative Responsibilities

"A colleague, not too long ago, asked me if my training had prepared me well enough to start my surgical practice. I responded that my surgical training was excellent preparation for me to deal with the variety of cases and emergencies in my practice. But the day I started to feel unprepared was the day when I took academic responsibilities. These responsibilities included; teaching, writing manuscripts and grants, trying to do research and administrative duties such as hospital committees, conflict resolution, managing financial budgets and fundraising."

Choices

Do you need to work in an academic center? Do you need to write manuscripts? Do you need to do research, to sit on several hospital committees or to hold the headship of a division or department? It appears that there is a trend in the last 10 years that a new recruit or a new person in the role of headship will be required to excel at everything. A person must have excellent clinical skills, be able to contribute to multiplc publications, be an active grants holder, have a strong research background, excellent managerial skills, etc. I call it "the super doctor." It creates an amazing amount of stress to succeed, to be better, and to have recognition. I am not too sure it is a healthy approach and I see it as a "distortion" of our profession. One sure fact is, that as you get older and more experienced in your practice, you become more involved in academic endeavors, in clinical research, in writing manuscripts, in teaching residents, in hospital committees and perhaps you become interested in a headship. You may be drawn to any of the aspects of an academic practice. You may feel the lure of it because you enjoy it and get fulfillment from it, recognition and rewards, but ultimately, remember, you need to prioritize your time. It is very difficult to be all things to everyone. There are only so many hours in a week, and your family – along with yourself - also deserve time.

The decision to pursue academic medicine (working in a university teaching hospital as a clinician or a scientist) is shared by a growing number of doctors across Canada. The vast majority of doctors choose private practice and it certainly applies to the majority of family practice physicians and a large proportion of specialty physicians across Canada. In a private practice setup, there will be much less opportunity for academic work such as teaching trainees, writing manuscripts or grants, and very few opportunities for research except perhaps for some clinical practice review. All aspects of hospital administrative duties will also be minimal and your administrative skills will be mainly about running your practice. For this type of practice setup and approach, learning the "ropes" on the job, so to speak, will be fairly easy. Academic and hospital administrative responsibilities are different, but can be blurred for many of us. It can create conflicts between the university and the hospital about who has the responsibility for what and to whom a physician reports to. I will try to explain the differences and the pros and cons. I will also include the difficulties and the benefits of this intricate relationship.

Academic Responsibilities

What is academic medicine? It is not just about clinics, surgeries, procedures, but also about advancing medical science through clinical, translational and basic research. This also includes developing the next generation of physicians, scientists and health care professionals. An academic medical center is comprised of multiple missions, numerous constituents including affiliated universities, various funding sources and multiple priorities, all of which converge to form an unprecedented degree of complexity.

Working in an academic university hospital setting brings a completely new set of responsibilities and duties, which I certainly did not feel I was prepared for. I had to adapt to change, not only in my field of practice, but in relation to the hospital organization and governance, in changing allied health professional's scope of practice and learning new leadership skills. Additionally, I took on new responsibilities such as the role of being the teaching director, program manager and division headship. Along the way, I gained access to many teaching tools and courses that I will bring forward in this chapter.

The experience of physicians in academic medical centres is filled with stimulation and stress, responsibility and reward, change, contradiction, and funding difficulties. Today's health care environment heavily influences the practice, the environment and the personal lives of physicians.

Leaders of academic medical centres, or departments, or divisions are traditionally selected from other organizations on the basis of their clinical and scientific recognition and because they are excellent scientists and clinicians. They are not necessarily chosen because of their good organizational skills or time management skills. It is a daunting task in today's health care system to be a leader and/or a department or division chief. Given the political and social stresses related to organizational complexity, fiscal constraints, diversity and generational differences, current leaders in academic medical centres need more than just clinical and scientific experience. Leaders must have their own views and approaches regarding issues such as promoting change, working with teams and managing conflicts constructively. Leaders must be able and willing to listen to others. I watched an interview with Larry King a few years ago and he commented "I did not learn anything by talking but I did by listening."

Leaders should understand that addressing cultures might be a key factor in improving workplace health, by reducing or eliminating workplace stress. Even in the best situation, changing organizational culture is challenging. In general, organizational culture is enduring despite attempts that may be made to change this. The key to success for any organizational culture is to effectively communicate a compelling vision of the future. Effective communication in healthcare will be explored further in Chapter 21.

Hiring expectations and academic advancements may be at odds with one another. Many clinical departments will structure staff contracts around clinical expectations with the understanding that the clinical faculty will figure out a way to be academically productive within the context of their clinical responsibilities. Such an understanding by the institution may not always be clear to the clinician. Add to this the possibility of changes in the future and discontent for all parties is inevitable. Recruitment should be viewed as an act of trust, but this is not always the case for multiple reasons. The individual, department chairs, potential mentors and collaborators should discuss the expectations and commitment involved in all work agreements. Please refer to the chapter on "The use of a lawyer". A document should be written that clearly states the work agreement outlining an explicit understanding of expectations for all parties.

Recruiting and retaining valuable faculty increasingly depends on creating an environment in which individuals can build satisfying careers without having to choose between personal and professional success. The most difficult challenge to the institution, the department, and/or the division, is to help individual doctors sustain their goals and expectations, to ensure that individuals have a manageable workload, equitable compensation and sufficient resources to fulfill his/her work.

Promotion and tenure throughout academic medicine has been a struggle; academic medical centres tend to promise less and less when it comes to tenure and in fact, some centres/universities have cancelled tenure altogether mainly for financial reasons, while others guarantee very little financial rewards.

In the 90's, the establishment of multidisciplinary centres has extended the trend toward placing the basic researcher and the translational researcher in a clinical context. Many centres have brought together clinical and research doctors as collaborators with the goal of applying research findings to clinical practice. The association of clinical, translational and basic research under one program model has led to many research advancements and clinical care improvements. Research physicians who became part of academic medical clinical programs have focused on increasing knowledge and technologies for the welfare of patients although they do not necessarily care for patients. Nonetheless, the same intensity and commitment is present. Since part of the challenge facing academic medical centres derives from the changes in medical practice, health care delivery and funding opportunities for research, the consequences of these changes have led to increased pressure on physicians, particularly those in academic medical centres where teaching and scholarly activities are important criteria for promotion in academic rank. Financial challenges affect physicians in academic medical centres as well as those in private practice and those who practice in hospital.

Mentoring is an activity that doctors engage in out of gratification rather than necessity. There are no consequences for being a negative role model and mentor. But mentoring is a professional responsibility that medical schools should support and recognize as a core academic responsibility. Trainees and young doctors require mentoring and appreciated it, and recognize the great benefits it can bring to anyone's career. Many residents and young doctors are not obtaining effective mentoring and most senior doctors are having difficulty adapting to the challenges and needs that mentoring represents. I was certainly not prepared to be a mentor. But being a mentor provided me with the opportunity to improve my communication skills, to take on new challenges and to better understand the changing medical field.

As a new pediatric cardiac surgeon, I felt privileged to work in a well-known Children's Hospital. I also aligned myself with the hospital's vision and mission. For better or worse, the hospital was my life. This is how I defined myself. It appears that younger trainees today have less faith and sense of belonging to an organization. While committed to their profession, they are rejecting the pursuit of their career through personal sacrifice. It certainly does not mean that they are any less professional or caring for their patients. Times have simply changed.

Finally, how can we hope to care for others if we, ourselves, are crippled by ill health, tired, burned-out, stressed-out, overworked or resentful? As a senior colleague and good friend told me, academic responsibilities and administrative duties may be a normal evolution of our career in an academic health center, partly ego, partly interest, partly commitment to the institution, partly pressure from colleagues but ultimately it is a choice that any doctor makes for themselves. In my friend's experience and mine, doctors like the changes that academic responsibilities and administrative duties bring to a busy clinical practice. The challenges bring a new level of interest, learning and growth. But I would concur with my friend, that no one can do it all, and we have to think about our personal health. My friend's advice, as well as mine, is to stop once a year, review the year and plan for the next year with an eye on where you want to be in 5 years. It is not only an excellent idea, but an obligation to yourself and family. With this planning you will have an excellent career path and also maintain the life balance for your family.

Hospital Administrative Responsibilities

The Canadian Healthcare System has faced multiple challenges over the years and many changes in the health care industry in the past two decades have mandated the re-thinking of the delivery of services to achieve both quality care and financial stability. As the population ages, people are living longer—often as a direct result of improved health care technology and procedures—and the number of patients is dramatically increasing. At the same time, the need for physical space and improved wait times is becoming critical. The expense of updating technology, the costs for both equipment and in acquiring resources for training, is continually spiraling upwards. In short, health care providers are being asked to serve more patients with fewer resources and with smaller budgets. But we have finite resources and the need for improved health care service delivery is inevitable.

We all understand that during the course of our training and in our practice, whatever practice we are in, there will be ongoing changes. We will have to adapt to new patient care treatments and delivery, to evolving technology (which in some instances may entirely modify our current model of care approach), to different organizational structure and governance, to changes in professional roles such as nurse practitioners, hospitalists, physician assistants and more. This may have to be done with fewer resources than ever before.

Quality Assurance and Management Tools

There are several tools available through books, courses, on-line learning and continuous medical education that will be helpful for you if you consider taking up hospital administrative responsibilities.

These tools are:

a) Quality Assurance
b) Quality Improvement
c) Risk Management
d) Operational Research
e) Six Sigma tool
f) Lean Thinking
g) Six Sigma Lean Thinking
h) Toyota Lean process and Lean Thinking
i) Lean Thinking Courses: The Leading Edge
j) Physician Management Institute (PMI) Leadership courses
k) Master Business Administration (MBA), part-time courses, full-time courses, on-line courses

I will provide information on these different tools so you can understand what they mean, how they are used and how you can access them.

a) Quality Assurance

Hospital's initial quality assurance programs were applied in the healthcare setting before risk management programs. Hospital committees consisting of medical staff leaders and nursing supervisory personnel dealt with quality-of-care, physician, or nursing problems on an individual basis. The concept includes the assessment or evaluation of the quality of care, identification of problems or shortcomings in the delivery of care, designing activities to overcome these deficiencies and follow-up monitoring to ensure effectiveness of corrective steps.

b) Quality Improvement:

Health Organizations and programs intend to assure and/or improve the quality of care through Quality improvement (QI) which consists of systematic and continuous actions that lead to measurable improvement in the delivery of health care services and the targeted outcomes. The concept includes the assessment or evaluation of the quality of care, identification of problems in the delivery of care, designing solutions to overcome these deficiencies and follow-up monitoring to ensure effectiveness of corrective steps.

Quality Improvement activities include the following:

1. Establishing specific quality-related goals to measure the organization's processes and outcomes
2. Administering programs that focus on improved outcomes of patient care or healthcare delivery systems
3. Providing consultative services to departments and services in the organization to assist them in achieving accreditation and organizational compliance in quality and performance improvement activities
4. Identifying opportunities for continuous improvement
5. Participating in root-cause analyses of events and designing solutions/systems to implement improvements

6. Evaluating patients satisfaction and initiating performance improvement activities based on the findings

Quality Improvement works as systems and processes such as define an issue/questions/problems, determine the current problem in the process by using data information, determine the requirement and steps for change through data analysis, create a plan for implementation, apply the improvement or changes, measure and document the effectiveness of the improvement and establish new standards.

c) Risk Management:

Risk management is the preventative process for managing risks. Risk management involves identifying risks, strategizing ways to avoid or mitigate those risks and developing a contingency plan in case risks cannot be prevented or avoided and understanding how risks can impact patients. Also, health care risk management is ensuring the continuity of patient care. This means that hospitals and doctors' offices have to consider how risks can be managed in a way that does not disrupt a patient's treatment .

Risk Management involves the measurement of the gap between the standard of care and the actual delivery of care. If there is a gap, it creates a liability/risk whatever it is in the delivery of care to the patient, the patient him/herself, the staff, the processes and/or the organization.

In the past, quality improvement and risk management functions were often dealt with separately in healthcare organizations. Individuals responsible for each function had different lines of reporting according to the organizational structure.

Today, quality improvement and risk management efforts in healthcare organizations are rallying behind patient safety and work together more effectively, and efficiently to ensure that their organizations deliver safe and high-quality patient care.

d) Operational Research:

Operational Research is the study of improving operations and decisions through the use of quantitative techniques such as optimization, probability, statistics, computer modeling, simulation, queuing theory, economics and mathematics.

The Operational Research to Modeling is to define the problem and/or ask the question/s, and gather the data. A mathematical model to represent the problem is formulated. A computer-based procedure program is developed and then used to derive solutions to the model and to simulate multiple scenarios representing solutions. The model and scenarios are refined and tested for their validity and the chosen scenarios that represent best the solutions to the problem/question can be implemented.

Since its first application during World War II for convoys organization, the Science of Operations Research has grown to a variety of applications in healthcare including; staff allocation, emergency disaster management, telecommunications, finance, business modeling, medicine, logistic/transportation, supply/demand chain management, inventory management, manufacturing, engineering, economics and more. It equally uses analytical and simulation models to describe, explain, predict and control systems, support the decision making of professionals and managers for resource allocation and applies methods of probability, simulation, optimization, differential equations, statistics and systems analysis.

e) Six Sigma Tool:

Defects or errors in any business carry an associated expense, but medical errors also carry significant human costs. And the IOM report estimates that medical errors cost the nation (USA) approximately $37.6 billion each year, with roughly $17 billion of those costs associated with incorrect or untimely diagnosis, wrong medication, delays in management, departure from professional standards, incorrect billing, etc. Errors in healthcare can result in part, from poorly designed complex systems.

For a multitude of reasons, improving healthcare quality is paramount. Communities today are not only demanding access to the best technology and treatment available, but also asking for assurances that medical encounters will be both safe and effective. In this competitive, quality-driven and cost-conscious environment, one of the most effective solutions healthcare professionals have found to this problem has been to adopt the Six Sigma methodology and other related change management techniques. Mounting evidence illustrates that using Six Sigma to design and refine patient care processes eliminates the need to retrace steps, correct reporting errors, re-do examinations or preventable errors in admission, and to have to reschedule appointments. Such redundancies and waste, are both costly in financial terms, as well as can potentially cause discomfort and dissatisfaction for the patient.

Six Sigma focuses first on reducing process variation and then on improving the capability of that process. Six Sigma is a systematic method for improving the output of the organization, by enhancing the quality of the system and processes. This is done by preventing mistakes, solving problems, managing change and monitoring long-term performance in quantitative terms so that any incipient problems are detected before they become a real concern. It uses a compilation of management practices to achieve its specified goal. Some of these practices are based on statistics, but the central idea behind Six Sigma is that if you can measure how many "defects" you have in a process, you can systematically figure out how to eliminate those "defects" and get as close to "zero defects" as possible.

f) Lean Thinking:

Lean Thinking is a radical new way to think about how to organize human activities in order to deliver more benefits to society and value to individuals while eliminating waste. "Lean thinking" originated in the automotive sector and is now at the forefront of all industries. It is especially valuable in the health care realm, where budgets are perennially tight and thus, efficiency is vital. "Lean" gives the health care industry not just a set of tools to use and procedures to follow, but a comprehensive and integrated thought process, culture and system of benefits.

The term "Lean Thinking" originated from Taiichi Ohno and was studied by James P. Womack and Daniel T. Jones to capture the essence of the Toyota Production System. Lean thinking is a way of assessing any activity and seeing the waste inadvertently generated by the way the process is organized.

This system focus on optimizing resources and delivering efficiencies through identifying and eliminating wasteful steps or activities in core processes, by analyzing any process and then determining its value by breaking the process up into 'value added steps' and 'non-value added steps'.

This is then further analyzed, with each step that is a non-value added step being completely eradicated from the process. This leaves a process that is made up purely of value added steps, so the entire process is completely efficient.

Lean Thinking views any activity that does not create value for the customer/patient to be wasteful, and thus should be removed or changed. If you trained every person to identify the time and effort wasted at their own job, the result in the workplace would be more value with less expense. At the same time, developing the employee's confidence, competence and ability to work with others. As healthcare costs continue to soar and systems are increasingly required to deliver better care to more people on smaller budgets, the promise of the "Lean Thinking" Methodology is compelling.

g) Six Sigma Lean Thinking:

Six Sigma Lean Thinking is an error reduction methodology which eliminates the root causes of defects and mistakes in a process. The fundamental objective of the Six Sigma methodology is the implementation of a measurement-based strategy that focuses on process improvement and variation reduction. The term combines two quality improvement approaches: Lean and Six Sigma. Lean methodologies focus on eliminating waste and streamlining processes. Six Sigma projects involve increasing quality and yield while reducing defects and variation. Six Sigma Lean Thinking methodologies can substantially improve the performance of many healthcare processes. Six Sigma Lean Thinking in healthcare projects have been used and have delivered substantial gains in projects such as:

1. Emergency Department patient flow and cycle time
2. Operating room patient flow and cycle time
3. Laboratory and Radiology cycle time
4. Patient wait time and wait list
5. Inventory assessment and ordering
6. Billing, coding and reimbursement
7. Medication Administration (antibiotics, chemotherapy)

Identify the benefits up front when creating their project charter. It encourages sustainable, long-lasting improvement, through its focus on setting up measurement systems and then consistently tracking and feeding back performance.

h) Toyota Lean Process and Lean Thinking:

Lean manufacturing was developed by the Japanese automotive industry, with a lead from Toyota and utilising the Toyota Production System (TPS). The concept of lean thinking was introduced to the Western world in 1991 by the book *"The Machine That Changed the World"* written by Womack, Jones, and Roos. Lean is a philosophy that seeks to eliminate waste in all aspects of a firm's production activities including: human relations, vendor relations, technology, and the management of materials and inventory.

The end result of applying the TPS to all areas of business comes from a five-step process: defining customer value, defining value stream, making it "flow", "Pulling" from the customer and back and striving for excellence. There is much more to learn about the Toyota Lean Process and additional information is provided in the references section of this chapter.

i) Lean Thinking Courses:

You can learn about the philosophy and methodology of Lean Thinking through books, presentations and courses. Several academic organizations and Universities across Canada provide learning courses on Lean Thinking.

Specifically, The Leading Edge Group is an international company that specializes in the area of change and continuous improvement by providing consultancy, corporate training and education to companies across multiple sectors. They also work in partnership with a number of academic and professional organizations internationally to provide accredited Lean certification programs. The Leading Edge Group is a leader in the provision of Lean and Six Sigma facilitation and change management delivering strategic and operational improvements for organizations by enhancing customer service, increasing quality and efficiencies and optimizing resources of the Lean education. They have also been collaborating with the University of Ontario Institute of Technology's (UOIT) Management Development Centre (UOIT-MDC) to offer online as well as education and training programs to organizations and individuals across Canada, such as the Hospital For Sick Children.

j) Physicians Management Institute (PMI) physician leadership courses:

PMI (Physicians Management Institute) physician leadership courses have three goals*("PMI Physician Leadership Courses")*. These objectives include; training which is specifically developed for physicians, is accredited by the Royal College of Physicians and Surgeons of Canada and the College of Family Physicians of Canada and is highly interactive. These healthcare leadership courses are available in-group session enrolment course formats, delivered in-house or available online.

If you choose a group session PMI leadership course, you will learn alongside colleagues from across the country and share experiences unique to your profession in a collaborative environment. In-house PMI physician leadership courses bring customized learning to your workplace, event, conference or annual meeting. The CMA will work with your organization to determine a physician leadership course that will be most useful to your audience. Training can be offered to physicians alone or to interdisciplinary teams that include physicians, other providers, managers and administrators. You will improve effectiveness in your work environment, by ensuring that you and your colleagues develop a common language and leadership skill set, collaborate to address common issues, solve problems and share new insights.

Online PMI physician leadership courses are offered and accredited by CMA. Online training affords the opportunity to learn at a time and place most convenient to the participant, as well as the ability to connect with peers across the country without leaving your office or home.

CMA Physician Leadership Institute (PMI) courses available include:

- Developing and leading system improvement
- Dollars and sense: leadership in the delivery of cost-effective health care
- Engaging others
- Leadership strategies for sustainable physician engagement
- Leading change and innovation
- Management dynamics: getting the job done
- Managing people effectively
- Physician as coach
- Professionalism and ethics
- Quality Measurement for Leadership and Learning
- Self-awareness and effective leadership

- Strategic influence: advocacy, alliances and accountability
- Strategic planning: from vision to action
- Systems transformation: navigating complexity through dialogue
- Talent management: a strength-based approach to developing physician leadership

These courses were available as Physician Leadership Course 1 to 4 several times per year in different locations across Canada, but have now been expanded into several more focused courses.

k) Harvard School of Public Health:

Executive and Continuing Professional Education offers innovative courses for leaders in public health, health care and environmental health and safety. These courses are designed to provide participants with the tools they need to discover effective solutions to issues critical to their practice and their organization.

As a result of environmental and financial pressures, academic health centres in Canada are experiencing major changes. In addition to the restructuring of the clinical services, academic centres are being challenged to sustain their academic missions and priorities. This challenge is a result of the continued financial constraints that the centres are facing. In order to tackle these challenges, institutions need physicians in administrative positions at all levels, who can provide leadership and managerial initiatives.

The Harvard Executive and Continuing Professional Education Program aspire to bring together physicians (these physicians are in administrative positions in academic health centres) for two weeks of intensive and systematic study. This study period includes reviewing the critical leadership and management issues which physicians encounter in administrative positions at academic health centres.

You can consult the applicable websites in the references section of this chapter for more information.

Tools Information Summary

While quality improvement and risk management methods can deliver improved efficiency and better quality products, the question of how applicable these methods are to healthcare can sometimes cause debate.

On the one hand, these large-scale industrial-type improvement methods can bring efficiency and effectiveness to healthcare. On the other hand, some may say that, people are not cars, a product or an assembly line and that the simplistic adoption of industry improvement process tools will only exacerbate the existing difficulties of delivering uniform, high-quality care with tight resources to populations whose expectations continue to rise.

Six Sigma is an ethos to decrease errors. Six Sigma has gained attention in healthcare organizations for being able to overcome challenges related to quality and cost.

Previously, healthcare organizations were frustrated by earlier efforts to improve organizational effectiveness. Any incremental gains they were experiencing were quickly offset by technology costs, an increase in patient population, incremental workforce shortage and offsets of other poorly performing departments.

Organizations do not have to choose between quality of care and saving money. Six Sigma and Lean Thinking projects improve care and reduce the likelihood of patient safety errors. It is grounded in measurement and statistical analysis, helping healthcare managers to develop plans of action that are backed up by data. It puts quality assurance and improvement tools in the hands of healthcare managers, who are familiar with the inefficiencies and hazards facing their patients, so that they can support organizational change.

In conclusion, the tools described in this chapter can be very beneficial for efficient health care delivery. A physician interested in the administration of his service, his organization, his university and government, should learn more about these tools and have a keen understanding of their principles, in order to provide meaningful interactions in the decision making process of health care delivery, it is essential to educate oneself in methods for continuous improvement. Remember however, that none of these tools will replace a good dose of common sense and judgment.

REFERENCES

Books:

1. H.A. Tata,
   Operations Research: an introduction
   Prentice Hall, PA, September 2010
2. W.A. Sollecito, J.K. Johnson,
   Continuous Quality Improvement in Health Care
   Jones & Bartlett Learning, Burlington, MA, September 2011
3. P.L. Shaw, C. Elliott,
   Quality and Performance Improvement in Healthcare: a tool for programmed learning
   American Health Information Management Association, March 2012
4. T. Cole, T.J. Goodrich,
   Faculty Health in Academic Medicine: Physicians, Scientists, and the pressure of success
   Humana Press, Totowa, New Jersey, December 2008
5. J.G. Lobas,
   Leadership in academic medicine: capabilities and conditions for organizational success
   American Jour. of Med.; 119, 617-621, 2006
6. W. A. Anderson, M. Grayson, D. Newton et al,
   Why do faculty leave academic medicine?
   Jour. of General Int. Med.,; 18 (Suppl.1,), 99-100, 2003
7. L.H. Pololi, S. Knight,
   Mentoring faculty in academic medicine
   Jour. of Gen. Int. Med.; 20, 866-870, 2005

8. S. Brown, R.B. Gunderman,
   Enhancing the professional fulfillment of physicians
   Academic Medicine; 81: 6, 577-582, 2006
9. D.G. Kirtch, R.K. Grigsby, W.W. Zolco,
   Reinventing the Academic Health Center
   Academic Medicine; 80 (11), 980-989, 2005
10. L.P. Howell, G. Servis, A. Bonham,
    Multigenerational challenges in academic medicine.
    Academic Medicine; 80, 527-532, 2005
11. M.L. George, J Maxey, D.T.Rowlands,
    The Lean Six Sigma Pocket Tool Book: a quick reference guide to 70 tools for improving quality and speed
    Mc-Graw-Hill, N.Y., 2004
12. J.P.Womack, D.T.Jones,
    Lean Thinking
    Simon & Schuster Inc, March 2003
13. M. Baker, I.Taylor,
    Making Hospitals Work
    Lean Enterprise Academy, Cambridge, MA, July 2005
14. J.P.Womack, D.T.Jones,
    Lean Thinking: banish waste and create wealth in your corporation
    Free Press, N.Y., 2006
15. J. Shook, C.Marchwinski,
    Lean Lexicon
    Lean Enterprise Institute, Cambridge, MA, February 2014
16. J.Liker,
    The Toyota Way
    Mc-Graw Hill, N.Y., January 2004

WEBSITES:

1. D.G..Kirsch: Culture and the courage to change 2007
   www.aamc.org
2. WHO: Quality of Care 2006: a process for making strategic choices in health systems
   http://www.who.int/management/quality/assurance/QualityCare_B.Def.pdf
3. Canadian Health Care Risk management Network
   www.chrmn.ca
4. Practice Management Modules in setting-up an office
   www.cma.ca/pmc
5. Physicians Leadership Courses (PMI)
   https://www.cma.ca/En/Pages/pmi-physician-leadership-courses.aspx
6. What is Six Sigma?
   http://www.isixsigma.com
   http://healthcare.isixsigma.com/library/content
7. Lean Thinking
   www.leadingedgegroup.com
8. MBA Studies
   www.mbastudies.com/mba/canada/online
   www.canadian-universities.net/mba/index.html
   https://ecpe.sph.harvard.edu/programs.cfm
   www.executive-bentley.edu

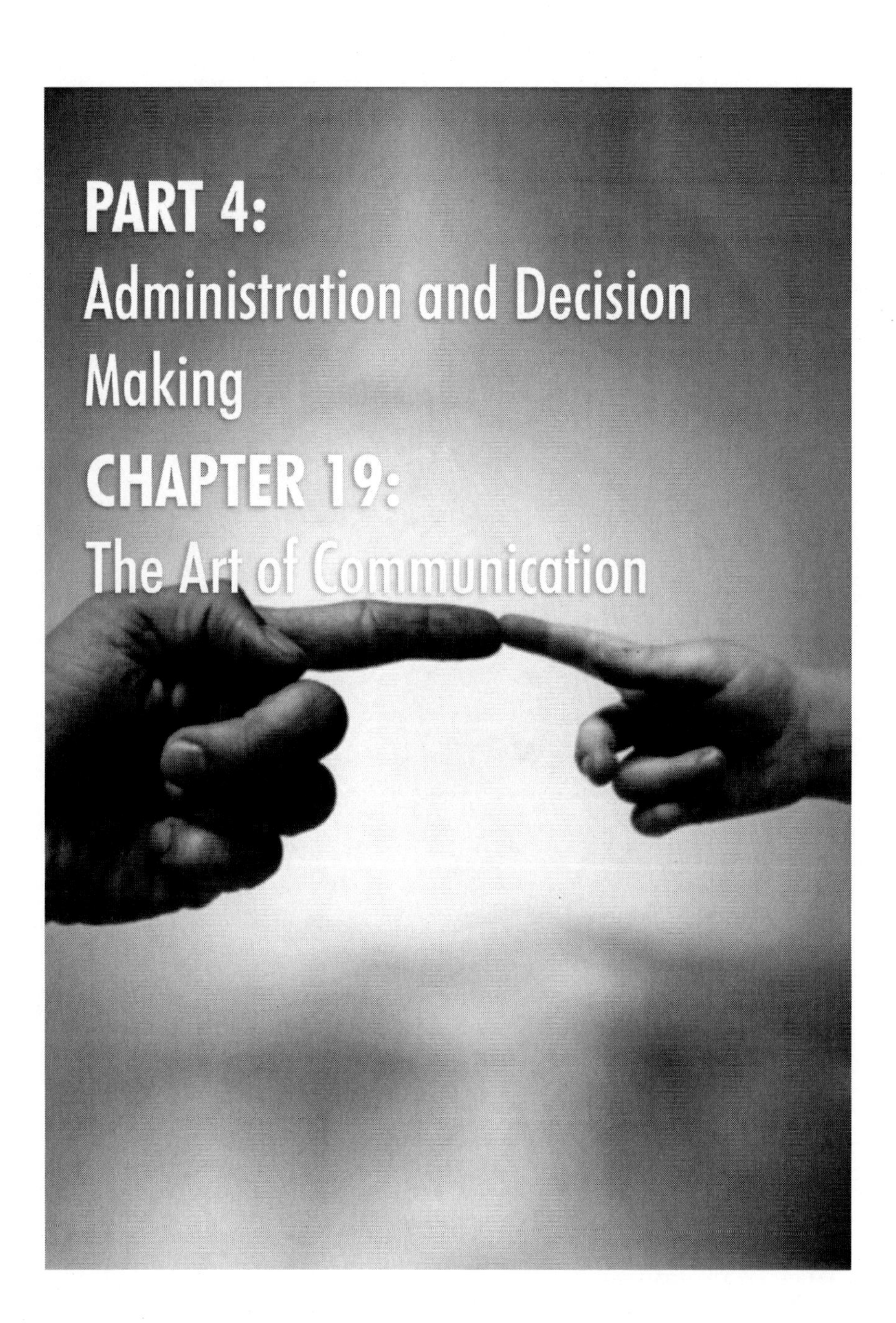

PART 4: Administration and Decision Making

CHAPTER 19: The Art of Communication

During my residency, I had three major aggressive interactions with another resident where I was angry and frustrated. I thought nothing at the time other than standing my ground in a difficult situation. When I started my practice it did not occur to me that medical colleagues could be unprofessional, lacking respect and collegiality in their interactions with each other. Fortunately, my senior mentor was a master at public speaking, at being controlled and mature and diffusing difficult situations. I was learning from the right person. Believe me, I had a lot to learn and along the way many individuals, skilled in the art of communication, helped me to be a better communicator. Chief among these was a group of nurses with advanced training who seemed to communicate with patients, families and health professionals more effectively.

While I have enlisted the support of experts in other chapters to inform, guide and validate my work I have used 2 healthcare professionals, Ms. Dori Van Stolk RN, BScN, MH, CEC and Ms. Susan LeBlanc BScN, MHSc. with expertise in the art of communication, to write this chapter. What I have come to understand is my very concrete and linear way of thinking does not always work well in communicating with others. As a surgeon, I realized that many people do not function in the way I do. I gradually learned that I needed to acquire more skills. Acquiring the skills led me to learn about myself overall and how to communicate more effectively with others. It is my hope and intention, in using these two experts, to expand your awareness of the shift we as physicians need to make in our understanding of what is involved in effective communication.

Communication is a basic need. It is a lived experience. In the workplace and in all areas of life it is an essential and central part of our being. The importance of communication, specifically effective communication, is increasingly recognized as a component of quality health care. Patient-centered communication standards are an established component of Hospital Accreditation (Commission, 2010). In relation to medical professionalism, improving physician communication enhances patient's welfare, social justice and healthcare quality (ABIM Foundation, 2002). What makes communication effective however, remains elusive and poor communication the fault or limitation of the other person. Since the 1960's it has been suggested that the ability to relate to another is central to effective communication and effective communication is essential to relating to another. It is a transactional, relational process that is complex and multilayered (Barnlund, 1970; Kilpatrick, 1961).

Healthcare is a relational industry. There are few professions that have the privilege of caring for patients and families in such an intimate way, involved in a time of their lives when they are most vulnerable. It would seem that communicating and relating effectively with each other would come easily and be a given skill that physicians, nurses and others in the industry come by naturally or at the very least have had some training in. In regards to physicians this is not the case. Much of physician and surgeon education is focused on becoming the "Medical Expert" and very little in the undergraduate and post-graduate curricula focus on the "Communicator" and "Collaborator" competencies (CanMEDS, 2015). So while the importance of effective communication is recognized and even required there is very little supporting the skill development of physicians to be able to communicate and relate effectively. As physicians and surgeons continue to specialize in being the 'Medical Expert', spending years honing their knowledge, skill and expertise in their chosen field, it may be at the expense of the time and attention required to internationally develop deeper communication and collaboration skills.

Communication is more than 'speech making'. It is not just the words we say but also the intention behind them, the meaning we give them and the tone and body language that accompany it all. Many of the physical barriers to communication have all but disappeared in this technological age, however the psychological aspects of communication remain. Effective communication is a science but more importantly it is also an art, in the ability to understand one's self, in the ability to attune to others and in the ability to express one's self in an authentic, compassionate way. The science of interpersonal relationships, psychology, and neuroscience, can be learned through reading and learning new skills and habits. The art of connection, empathy and compassion must be practiced, sometimes re-learned and embodied as a mindset, rewiring deeply engrained belief structures. This chapter introduces considerations, strategies and tools to become more effective at relating to and communicating with other physicians, other professionals and patients/families.

Awareness of Perspective

"Everything that irritates us about others can lead us to an understanding about ourselves."

-Carl Jung

Above the line, below the line; high road, low road; learner vs. judger: these are all metaphors for how we behave in relationship with another human being. As humans we are meaning-making machines and view the world from our own perspective and perceptions. Perceptions are shaped by the vast amount of data we take in both internally and externally. Our internal environment is shaped by past experiences, family of origin, hopes and expectations. Our external environment is shaped by measureable and objective data. As humans we aim to make sense of this data, moment to moment, to determine how to behave, react and interpret the world around us. These 'mental models' become our internal operating system from where we function in our daily lives (Senge, 1999).

However, much of what we believe to be truths are actually 'stories' that we create in order to make sense of the data that we internalize and filter. These 'stories' are formed by a complex interplay of the data available to us, the feelings that emerge within us to drive our behavior, and the assumptions and judgments we make. This primitive interplay helps to keep us in a state of equilibrium, safety or status quo and helps us respond to perceived threats of present day. We might refer to it as the way we 'are'. It is a subconscious act and built from many years of arranging and rearranging our perceptions.

Several types of distortions can occur that keep us 'safe' in the sense of self we feel most comfortable in (Bushe, 2010).

Perceptual Distortions

- When we form our 'stories', they become our truth and reality. Once formed, it becomes more difficult to update or change our story. We resist allowing any new information that may shift our belief structure i.e. we don't update our operating system to function in a new way.
- We filter and choose what we pay attention to. We notice those events and situations that fuel our 'story' and subconsciously discount information that may challenge our 'story'.
- Human brains have a built in "negativity" bias (Hanson, 2013). We make up to 80,000 thoughts a day and 60,000 of those may be mostly negative. Our belief structure develops a "problem-reacting" stance and we pay more attention to perceived threats and assume the worst.
- Our stories tend to amplify differences and find fault in others for example: "He's self centred" and "She's arrogant". Whereas we blame external causes for our own misgivings for example: "They made me late" or "They forgot to remind me" or "They scheduled the OR slate wrong".
- Our stories are more often projections of ourselves. In other words, it is almost impossible to perceive something outside of ourselves that isn't wired inside somehow, somewhere.

The Ladder of Inference by Argyis (1990) describes clearly the process of sense making that we go through.

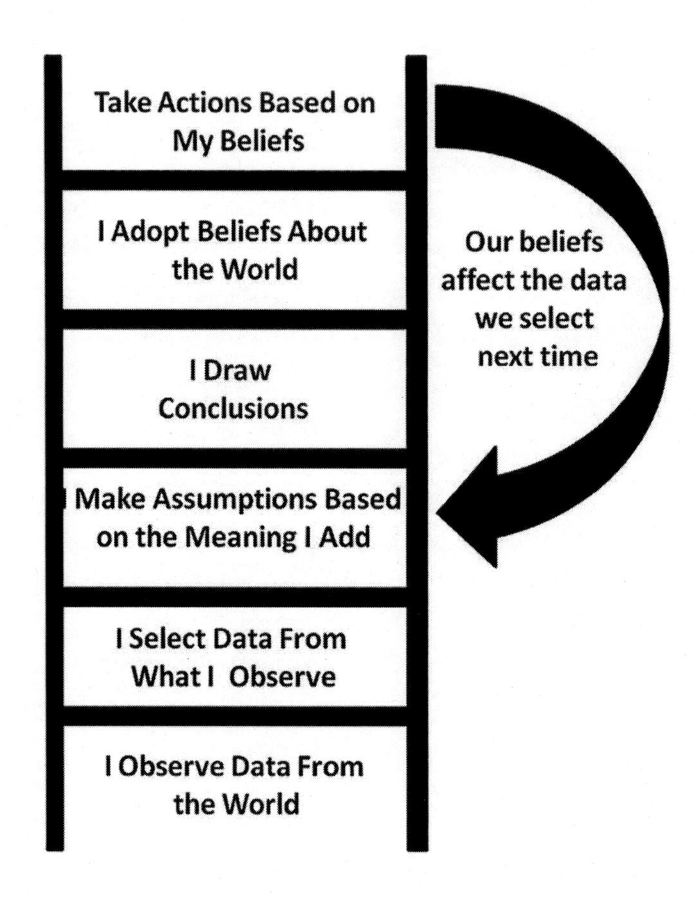

http://www.lopn.net/LOI.html
http://ed.ted.com/lessons/rethinking-thinking-trevor-maber

The Ladder draws our attention to how and when we make meaning of the data and experiences in our lives that become our perspective and inform our reaction or response in a given situation. All of this happens in a split second. Once we begin to be aware of how our perceptions are formed and how we respond, the work we need to do as meaning-making machines is to reshape our perceptions and muster up a more thoughtful response. Easy to say, harder to do!

The Interpersonal Gap

"The most basic and recurring problem in social life is the relation between what you intend and the impact of your actions on others."
-John Wallen

It is truly remarkable that anyone can actually have an honest, authentic conversation given our various perspectives, assumptions and stories. The 'Interpersonal Gap' (IP Gap), first described by Wallen in 1960 (Chinmaya & Vargo, 1979) is a model of human interaction illustrating how and why communication gaps happen. Effective interpersonal communication requires us to shift and change our story and seek to understand the perspective of others. When we fail to do so, the IP gap widens and confusion, conflict and disruption in the relationship occurs.

EXAMPLE 1

RN Perspective: Kathy's story

Around 9 pm on a Saturday Kathy is concerned about the status of a surgical patient. She calls the surgical fellow via the hospital operator. The surgeon arrives on to the unit, seems angry, raises his voice and mutters something about why he was called. He checks the patient, reviews the chart, writes orders and leaves abruptly. The nurse looks at the orders and realizes that she will need to call him again as the order she was calling about has not been completed.

The nurses all gather around the nurse at whom the comments were directed. Kathy is upset and doesn't understand why the surgeon was so angry. In addition she now needs to call him back and she is sure this will result in more conflict. She wonders if she should wait until the morning.

Surgical Fellow Perspective: Michael's story
Earlier in the day, Michael was called in for an emergency OR. He had been in the OR all day, and was exhausted. He was on his way home to have dinner when called back in. He arrived on the unit only to find that the surgical resident could have easily resolved the issue. He doesn't understand why the nurses," just don't follow the call process". Responding to the call has added to what he must get done before returning to the hospital early in the morning for surgery.

The Interpersonal Gap
Surgical Fellow: Michael's intention was to remind the nurses of the call process. He has strong values and beliefs related to following processes and 'hierarchy' of the call process. He feels this is one of the ways that the medical staff can organize the workload so that the doctors are at their best and patient's get the best care.

Nurse: Kathy felt disrespected and dismissed. She followed her gut instinct and was concerned about patient safety. She believes that patient safety is essential to giving the best care possible. She was always taught to work through differences in a respectful manner and yelling was not part of how she grew up resolving conflicts.

In order to narrow the interpersonal gap we need to bring clarity to our own internal experience and intention, and then turn to the other to seek to understand their internal experience and the impact of our intention. It requires a shift in our mindset from one of judging others and self, to a mindset of curiosity and learning.

The Learner vs. Judger Mindset

(Adams, 2014)

As meaning making machines, humans make judgments all the time and live in 'stories'. It's how tightly we hold on to our judgments that can be dangerous! When we're in judger mode, we judge others (blame) or judge ourselves (shame). This mode of operating impedes the ability to relate to and communicate effectively with another.

The more effective mindset to embrace is the learner mindset. Being in a learner mindset means being able to start or shift to a place of curiosity and seek to understand the other. When triggered, adopting curiosity, inquiry and seeking to understand self and others is the path to being able to communicate effectively. *Easy to say and much harder to do!*

EXAMPLE 1: Continued

Adopting a Learner Mindset

In the previous scenario Janet, one of the nurse's who observed this interaction, followed the surgical fellow out and asked to speak with him.

Upon further exploration she learned from him that he had been called into the hospital that morning for an emergency case. He had been in the OR all day, and was called back into the hospital earlier that evening. He was just on his way home when he received this call. He didn't understand why the nurses didn't call the resident first, before calling him. He was tired, and hadn't eaten yet. He had more work to do in preparation for a large case the next morning. He didn't understand why the call process wasn't followed since it had been established a year earlier. He finds it frustrating to work on something that is supposed to improve care and yet doesn't get implemented or communicated. The nurse thanked him for the conversation and went back to the unit.

On the unit, she asked the nurses what their understanding of the call process was. They had no idea of the call process. They were not regular nurses on the unit as there had been a merger of two units while one was being renovated. Having surgical patients on a unit of this nature was a bit unusual. They were concerned about the status of the patient so thought they were doing the right thing.

EXAMPLE 1: Continued

Bridging the Gap

It is clear to see from the nurse's and surgical fellow's stories that each was coming from a different perspective and impacting the other in such a way that limited his and her ability to communicate patient care needs effectively.

In adopting curiosity and approaching the surgical fellow and nurse to ask what their perspectives were Janet was demonstrating a Learner Mindset. She was able to identify the discrepancy or gap in perspective allowing for the development of a more thoughtful response that would acknowledge the common ground and meet the needs of those involved.

Impeding our ability to be in a learner mindset is the presence of triggers. Triggers are anything that provokes in us a feeling of threat and activates the stress response. They may occur many times a day and can be the result of:

- Clash of values
- Differences in beliefs
- Emotions
- Threat (real or perceived)
- Confidence

Once triggered, we strive to stay safe in our story. We make assumptions and judgments reinforcing our tightly held beliefs rather than adopting curiosity to learn and understand self and others.

When we are triggered, we have a physiological response that actually prevents us from being in a learner mindset and taking the learner path. We fight, freeze or flee. That damn, Amygdala! This is the stress syndrome, activating our sympathetic nervous system (Boyatzis, 2005). We then react based on emotion, because physiologically we've actually by-passed the frontal cortex, and we operate straight from the reptilian brain…damn Amygdala!

In order to switch from a judger mindset to a learner mindset and take a learner path we need to manage our response to triggers – the things that set us off.

Notice, Name and Tame:

A first step is to recognize when you are being triggered (Notice), what it is that is triggering you (Name), and then how to manage those triggers (Tame). Though typically the action or inaction of someone else triggers us the response belongs to us! It usually elicits a clash with one of our values, triggers a family of origin dynamic, or some other deep-rooted belief in our sense of self.

When we enter the 'stress syndrome' our brain basically shuts down our non-essential circuits. We are then less open, less flexible and less creative. Our brain loses its capacity to learn. We perceive things people say or do as threatening or negative, and more stress is aroused.

EXAMPLE TWO

It is a busy medical unit that also serves as the post surgical cardiac unit. The surgeon comes onto the ward and seems irritated that his patients are not all in the same area of the ward. The charge nurse intercedes and a conflict arises.

Judger path: *The charge nurse assumes that the surgeon is upset and mad at her. She takes it personally and thinks she did something wrong.*

Learner path: *The charge nurse takes a moment to observe her own internal experience. She is feeling a little anxious. Her heart is racing a bit. She takes a deep breath and wonders what might be going on for the surgeon. The charge nurse listens to the surgeon, and asks him what he is most upset about. He tells her and she acknowledges his position on the situation. She asks him what his* *understanding is of how the nursing assignments are made and what other considerations she needs to make related to staffing concerns.*

In taking the Learner Path, the charge nurse first noticed, then named and tamed her trigger. She adopted curiosity, as evidenced by asking open questions and sought to understand the surgeon's perspective.

Emotional Intelligence (EQ) and Emotional Competence

Emotional Intelligence is the term given to a group of abilities that enable human interactions to work effectively: the ability to be aware of one's own experience, the ability to be aware of others experience and the ability to effectively manage one's self in relationship. Understanding self and the role that self plays in any given interaction is the foundation of emotional competence. Emotional development begins very early in life as we experience attachments with the adults around us. We learn a range of coping strategies to deal with the stress of our emotional self in relationship with others. Coping strategies while relieving stress initially can also become engrained patterns of behavior that further the stress response and lead to resistant, repetitive behaviours that interfere with developing and maintaining relationships.

Becoming aware of and managing our internal experience or stress involves a complex interplay of the information we take in from all sources, the way we process and infer from it, and how we take action. Much like the process of communication itself. Boyzatsis (2005) suggests that the way to deal with our internal experience is through renewal responses including self-awareness, hope and compassion. Engaging in renewal responses decreases our stress response and allows for greater capacity to relate to another and reply with greater thoughtfulness. Emotional Intelligence is a growing field and the subject of many books, workshops and teachings that are worth exploring further.

EXAMPLE TWO: Continued

The Charge nurse in the above example practiced mindfulness to bring awareness to her own self about being anxious and engaging in a potential conflict. By deep breathing, she was able to calm herself down, and then be available to the 'other' (the surgeon) to be able to inquire about his experience. This is relational attunement and differentiation.

The Art of Knowing Self

"The heart is like a garden: it can grow compassion or fear, resentment or love. What seeds will you plant there?"

-Jack Kornfield

Engaging in creative arts or creative activities can facilitate the development of thoughtful more evolved responses through a greater understanding of self and identification of triggers. Creative arts of any sort serve as a channel for releasing, elevating and understanding our inner conflicts, fears and intentions as well as our hopes, aspirations and ideals. Creative activities can also help us to see the stories we may be telling our self and others.

Everyone is capable of developing a creative activity. Whether we choose to or not is another matter. There are no rules or techniques that must be followed or "natural" talent required. The key is to experiment and find a form that works for you. Intuitive painting, journaling, meditation, yoga, walking, cooking and playing a musical instrument are all forms of creative activities that allow you to develop insight and self-awareness. You may feel that your work is a form of creativity that brings you to a better understanding of yourself. If this is the case then develop a practice of awareness that acknowledges and integrates the insights you develop about yourself through your work. Integrating what you learn about yourself is an important aspect of developing self-awareness and having it influence your intentions and behaviours. In the words of Brene Brown (2015), "Creativity helps us to move things from the heady space of knowing to the heart work of practicing."

This follow through, if you will, of "knowing" into "practice" is akin to what science refers to as knowledge translation. The process of recognizing and translating our knowing of inner self into the way we interact with people and the world around us. It is in essence the deep work healthcare professionals need to do.

Rarely do we allow ourselves the time on a regular basis to engage in activities that enable us to see and move off of an "old story", to open to a greater understanding of what can be and evolve our ability to respond to others in a compassionate and authentic way. In the busy world that we exist in we consistently place developing the self and insight into the self far down the list of priorities. In fact, it is extremely important that we take time each day, even for 10 minutes to engage in some form of creative activity. The challenge for many is that the thought of personal insight through creativity generates anxiety. Consider this a process by which you develop the skills and ability to focus on learning about your self over time. Ten minutes a day translates into a significant amount of time over the years and if it provides a better state from which to function, you may be using your time more effectively in other endeavours. By not investing in learning about yourself, quality of care and your quality of life are put at risk.

Connection in the Age of Distraction

Our Busy World is Becoming Busier-Technology

"Technology gives us power, but it does not and cannot tell us how to use that power. Thanks to technology, we can instantly communicate across the world, but it still doesn't help us know what to say."

-Jonathan Sacks

We have now become enlightened to the keys to successful communication and relationships. It involves slowing down our internal world, noticing, observing self and relating to others, adopting practices that keep us in learner mode and responding in a more thoughtful way. In the fast paced environment of healthcare, it is imperative we find time and space to discover more about our own self. That doesn't mean plugging into the Internet, checking emails and believing we are connected. That would only be telling a story to our self that we are doing our best to try and keep up with everyone. As a society, we have entered the 'Age of Distraction" whereby the technology at our fingertips takes us further from relational attunement. Our healthcare world has also become more technologically savvy with the introduction of electronic health records. The next challenge we face in maintaining effective communication is to be skillful at interfacing with a computer screen and keyboard, all the while being focused and present to our patients. There is a tendency to turn our backs or type and talk, with the occasional 'uh-huh', or 'I see', or 'yes', without even looking at the patient. For the most part, patients view the electronic chart as beneficial to their overall care. Electronic health records can provide greater access to and sharing of information across all service providers potentially aiding in diagnosis and treatment. Addressing concerns about privacy and confidentiality, accuracy of information and appropriate sharing of information with key healthcare clinicians is imperative. It is an element of the work we do that speaks to developing trust with those we are in relationship with.

Here are a few tips to consider with E-charting:

- Be explicit and factual with your patient about your intentions when using the computer, "I am just going to pull up your chart on the screen here"; "I may be typing some notes as you talk, and then I will pause to make sure I have captured your information correctly."
- Let the patient know if your documentation will take longer than usual or expected.
- Make eye contact.
- Adjust your position so that you face your patient. If at all possible do not have your back to the patient.

- Review the data entered into the record with the patient so that he or she is clear on what you have documented.

We are still new to the era of electronic health records and much ahead remains unchartered. As inter-professional collaborative practice evolves and patient-centred care becomes the norm, we must all be accountable to an ongoing practice of self-awareness and developing the skills and ability to relate to and communicate effectively with those we serve.

"The newest computer can merely compound, at speed, the oldest problem in the relations between human beings, and in the end the communicator will be confronted with the old problem, of what to say and how to say it."
-Edward R. Murrow

Learning in Action

The bad news is there is no magic bullet for changing your ability to communicate effectively. The good news is there are many ways for you to explore and get started with developing your understanding of yourself and relating to others.

CHECK OUT

- Personal or Executive Coaching services.
- Lectures, seminars or workshops on interpersonal communication, relationships and leadership.
- Classes on creative activities such as yoga, intuitive painting, journaling, music, cooking, photography or pottery.
- Practicing meditation, yoga, mindfulness, breathing exercises.
- Physical exercise-even just a 20 minute walk per day.
- Take 1:1 counseling with a qualified practitioner who works with medical professionals.
- Find a mentor.

- Ask for feedback and find opportunities to be observed in practice to receive feedback about your communication patterns. Have a 360-feedback review done for learning purposes.
- What is available on the Internet? Use the search words: Communication, emotional intelligence, interpersonal relationship, leadership, transactional, awakening, self, insight, coaching, mentoring, personal growth, empowerment, healthcare, medicine, education, empathy and compassion.

RECOMMENDED READING

Change Your Questions, Change Your Life: 10 Powerful Tools for Life and Work
Marilee G. Adams

The Courage to Be Present
Buddhism, Psychotherapy, and the Awakening of Natural Wisdom
Karen Kissel Wegeh

Art Is a Way of Knowing
A Guide to Self-Knowledge and Spiritual Fulfillment through Creativity
Pat B. Allen

Being True to Life
Poetic Paths to Personal Growth
David Richo

The Book of Awesome
Neil Pasricha

Perfect Love, Imperfect Relationships
Healing the Wound of the Heart
John Welwood

REFERENCES

1. ABIM Foundation, American Board of Internal Medicine, ACP-ASIM Foundation, American College of Physicians-American Society of Internal Medicine, European Federation of Internal Medicine. (2002). *Medical professional- ism in the new millennium: a physician charter.* Ann Intern Med. (136) 243-246.
2. Argyris, C. (1990). *Overcoming Organizational Defenses: Fascilitating Organizational Learning.* Eds. (1)
3. Barnlund, D.C. (1970). *A Transactional Model of Communication.* In Akin et al (Eds), *Language Behavior: A Book of Readings in Communication* (pp. 43-61). The Hague: Mouton.
4. Boyatzis, R. & McKee, A. (2005). *Resonant leadership.* Harvard Business School Publishing, Boston, MA.
5. Brown, B. (2010). *The Gift of Imperfection: Let Go of Who You Think You Are Supposed to Be and Embrace Who You Are.* MN: Hazelden.
6. Bushe, G. (2010). *Clear Leadership.* Davies-Black, Boston, MA.
7. Chinmaya, A & Vargo, J.W. (1979). *Improving Communication: The Ideas of John Wallen. Canadian Counsellor,* 13(3), 152-156.
8. Commission, T.J. (2010). *New and revised standards and EP's for patient-centered communication, prepublication version.* Retrieved Nov 11, 2011, from http://www.imiaweb.org/uploads/pages/275.pdf
9. Frank, J. Snell, L. & Sherbino, J. Eds. (2015). CanMEDS 2015 Series IV. *Physician Competency Framework.* Royal College of Physicians and Surgeons. http://www.royalcollege.ca/portal/page/portal/rc/common/documents/canmeds/framework/canmeds2015_framework_series_IV_e.pdf
10. Goleman, D. (1995). *Emotional Intelligence.* Bantam Dell, New York, NY.

11. Goleman, D. (2013). *Focus*. HarperCollins Publishers, New York, NY
12. Hanson, R. (2009). *Buddha's Brain*. New Harbinger Publications Inc. Oakland, CA.
13. Hanson, R. (2013). *Hardwiring Happiness*. Random House LLC, New York, NY.
14. Kilpatrick, F. (1961). *Explorations in Transactional Psychology*. New York University Press. NY.
15. Miller, Sherod, Phyllis Miller, Elam W. Nunnally, and Daniel B. Wackman. *Talking and Listening Together: Couple Communication* I. Littleton, Colo.: Interpersonal Communication Programs, 1991. Print.
16. Senge, P. (1999). *The Dance of Change*. Doubleday, New York, NY.
17. Short, R. (1998). Learning in Relationship. Learning in Action Technologies

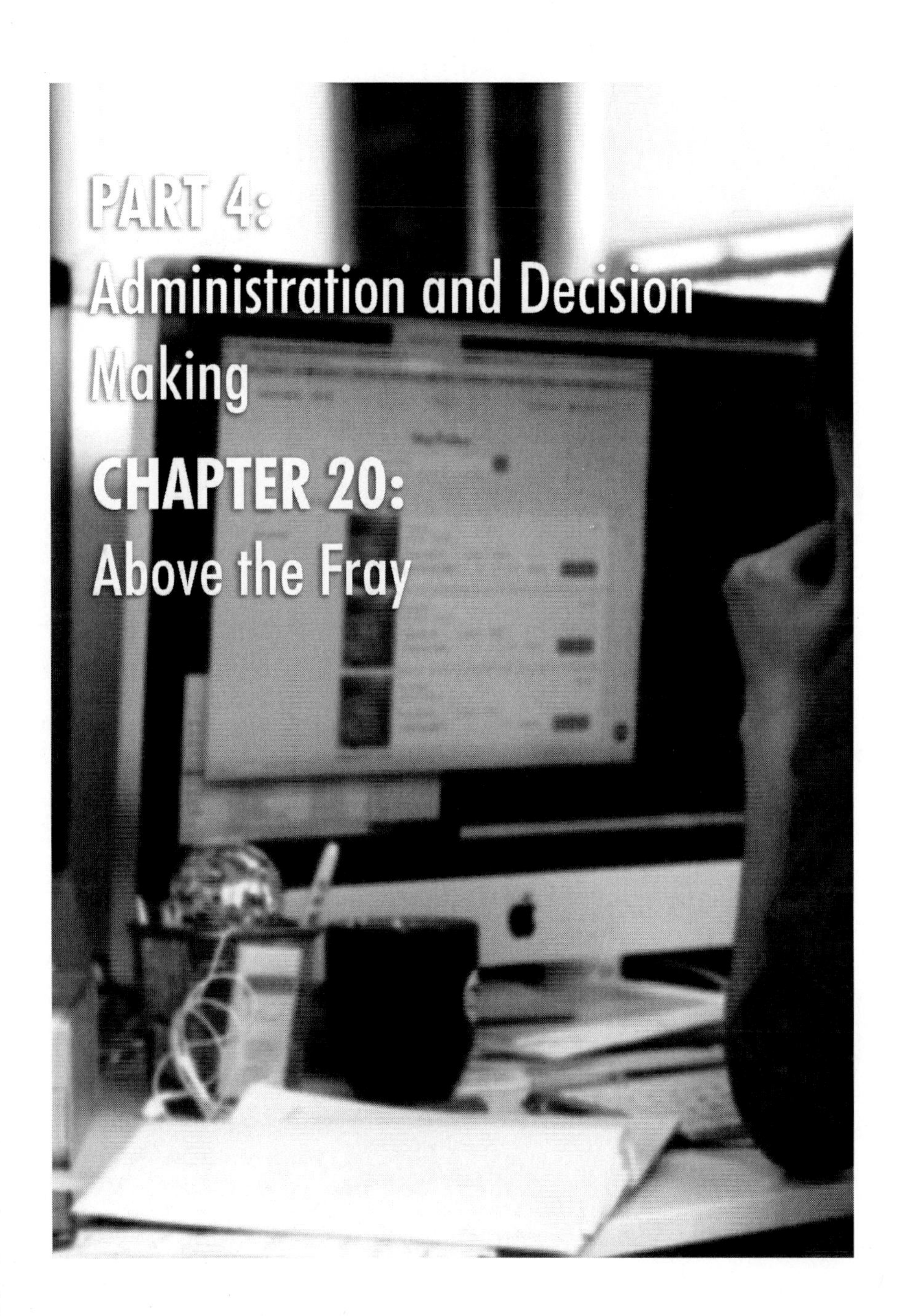

PART 4:
Administration and Decision Making

CHAPTER 20:
Above the Fray

Teamwork

As a resident in training, we learned to work as a team and it became part of our knowledge and skill set to practice our specialty. At the time of our training, we probably did not realize that this teamwork would be such a beneficial skill to have as we start our own practice. Team skills are very helpful, for the most part you will be working with others, unless you have a solo family practice. With a team you may have several players, there are many different personalities and everyone can be working in different roles. You should at all times, use respect, politeness and compromise when interacting and communicating with the members of your team. It only takes one difficult team member to spoil the atmosphere, cooperation and efforts of the team.

A team usually requires a leader to keep everyone on the same path with the same goals. The leader will make timely decisions, to review complicated issues and to address different behaviour (difficult personalities or people who are acting unprofessional).

Code of Conduct and Inappropriate Conduct

Academic health centers are not always a model of democracy; but we must be mindful of the demoralizing, disrespectful, and destructive consequences of the abuse of power. What constitutes the abuse of power? Power exercised hierarchically or not and impersonally, without consultation and the opportunity for dialogue. We recognize in medicine the ethical principal of colleagues, but academic health centers do not always show the same respect for their faculty members and doctors as employees.

Disruptive physician behavior does exist in our profession and some doctors act in a way that is disrespectful, unprofessional and toxic to the workplace environment. We wouldn't necessarily consider unprofessional behavior to be as offensive in the workplace as say sexual harassment, working under the influence of alcohol or drugs, or gross incompetence. However, in my experience, this nature of behavior can extend to a wide range – sometimes causing similarly catastrophic damage.

Proper physician behavior in the conduct of professional activities is expected from the medical profession and teaching professionalism is mandated by the Royal College. Medical professionalism does not have a precise definition, but encompasses a wide range of behaviors such as proper bedside manner, care for patients, respect for all members of the medical team, and proper adherence to the fundamentals of medical ethics in patient care and respect of hospital by-laws. Patients expect their physician to be professional while providing excellent clinical care. Patients can accept a physician's shortcomings (lack of communication, no smile, busyness, cold attitude), if their medical/surgical care is adequate; however, they should not have to settle for unprofessional behavior by their physicians such as misconduct, arrogance and poor communication, which can significantly affect the quality of care and patient safety issues. This poor professional conduct can lead to medico-legal issues.

Dishonorable behavior can take other facets, such as: imposing work on a colleague, refusing to see a patient, insults, inappropriate comments, anger, discrimination, harassment (sexual or otherwise) refusal to cooperate, refusal to follow protocols and spreading rumors. Jealousy exists among physicians and it can be related to perceptions, to another person's greater success, to another doctor having better medical or surgical skills.

Lack of support from colleagues, back stabbing for personal interests and being generally discriminated upon during a period of sickness can all occur. Disruptive behavior is seen with surgeons who are verbally abusive and throw instruments in the operating room, with physicians who make errors in clinical judgment due to impairment or other causes and physicians who intimidate the nursing/allied health staff (so much so, that the nurse and allied health staff are unwilling to question an inappropriate order or safety issue). Finally, addiction issues of any kind, falsifying documents and financial fraud could also occur.

"Mike is an excellent surgeon. Everybody respects him and knows that in difficult surgical circumstances, Mike can make miracles. But in these difficult surgical situations, Mike gets easily frustrated, agitated and throws instruments. The staff always accepted it, as it was temporary, until one day, in throwing an instrument, he injured the hand of the scrub nurse who then required surgery. He was disciplined."

"Peter is a pediatric urologist with many years of experience. He is surprised to get a request from an adult surgeon of another hospital to help him book a 10-year-old patient. He realizes that the child was referred to the adult surgeon by one of his pediatric colleagues who had not referred or consulted to Peter on the case. Ultimately, the child received appropriate care, however, it was a question of inappropriate referral of a pediatric case to an adult center that could have resulted in substandard care. The referring pediatric colleague demonstrated poor judgment and a lack of professional respect."

"Matt created an intolerable work environment for Shawn, through consistent disagreement, challenging openly his medical decisions, condescending attitude and refusing to acknowledge and support his career advancement. It led Matt to eventually leave the institution."

Could the medical profession be more prone to unprofessional conduct because of stress, long hours, bureaucratic administration and constraint of financial resources? In this era of changes, of rapidly evolving technology, of demanding patients, of financial reforms and shrinking income, it is a challenge to always be professional and polite. But again, many other professions encounter similar issues.

Many doctors have witnessed colleagues yelling or insulting staff, making inappropriate jokes, discriminating against colleagues or patients, spreading malicious rumors, refusing to cooperate with other health care personnel or refusing to follow the institution's rules. The least common, but one of the more disturbing behaviors, are cases of throwing objects, retaliation, substance abuse and physical violence.

It is very difficult to conduct a reliable and unbiased study about unprofessional behavior because doctors may not perceive themselves as having unprofessional behavior. They may not admit to it and certainly are less likely to report it in a study even if it is anonymous. Other doctors may report what they think is unprofessional behavior, sometimes not knowing the full context of the story. Finally, many behaviors are not seen as unprofessional, as the definition of inappropriate or unprofessional behavior is difficult to identify with accuracy. There will always be an element of subjectivity.

Addressing Inappropriate Behavior

Unprofessional and disruptive behavior by a physician must be dealt with according to the organization's medical staff rules of conduct. The primary goal of dealing with unprofessional behavior is to protect patients and staff, and ensure safe and appropriate clinical care. The second goal is to develop an appropriate remediation or recovery plan for the physician so that he or she has the chance to return to the practice of safe, appropriate, and professional medical care. Finally, the institution itself must think and prepare for possible medico-legal risks.

There are many different causes for unprofessional behavior. Some are more common than others, such as personal or family stress factors, financial difficulties, health problems, and substance abuse (drugs, alcohol). Serious inappropriate or disruptive behaviors require identifying the underlying problem: physical or mental health issues, financial problems or substance abuse. Appropriate planning for treatment or help must be established after careful consultation and counseling with the physician. Then if needed, there are many national and international treatment facilities and local resources that could help get the physician back on the path to a professional level of conduct. Reinstatement to practice, full or limited, should be the goal after successful completion of therapy/counseling and plan for a period of close monitoring over the physician's practice. This should be put in place and agreed upon by all involved parties. More serious cases of unprofessional behavior should be reported to the respective provincial college authority for its separate investigation and possible disciplinary action.

Physicians do have ideas about what needs to be done/improved for this issue. Some of the suggestions in the field have been to establish a framework for confronting/reporting disruptive behavior, strategies for discipline and steps to monitor the rehabilitation off work and back in the workplace. Despite the fact that health care organizations have these mechanisms in place, physicians in general have personal difficulties to report a colleague for all kinds of reasons. Some physicians in position of authority find it difficult to confront another colleague. Do not be surprised if unprofessional behavior does not always get addressed like it should.

It is important that every healthcare facility, academic medical center, and even physicians group practice have a well-developed medical staff code of conduct. You can check your respective provincial college authority for more information about their code of conduct program. The medical staff bylaws should clearly state the procedures for reviewing and investigating complaints against medical staff members and direct the proper process for disciplining and/or remediating a physician whose misconduct has been adequately substantiated. Also, for further reference, the American Medical Association (AMA) has a very well developed Model Medical Staff Code of Conduct *(ama-assn.org)* and it is an excellent document and blueprint for this sort of ethical dilemma.

In summary, unprofessional behavior or conduct by a physician or another colleague is a reality, so don't be unprepared. A colleague's unprofessional behavior can be directed at you. They may verbally attack you, spread malicious rumors and even affect your career. Not only can it be detrimental to you personally, but it can also be damaging to the physician themselves, to his/her family, to their quality of patient care and to the healthcare profession. An organized approach to investigating complaints against a physician must be provided by the health organization/hospital/university and by the provincial college authority with an obligation to protect patients and staff as well as provide due process for the physician. Ultimately, therapy/counseling/coaching will be in the best interest of the involved physician, so always take action where appropriate.

REFERENCES

Books:

1. Owen MacDonald,
   Disruptive Physician Behavior
   Group Publisher, QuantiaM, May, 2011

Websites:

1. http://policybase.cma.ca/dbtw-wpd/PolicyPDF/PD04-06.pdf
2. www.ama-assn.org/ama/pub/footer/code-conduct.page

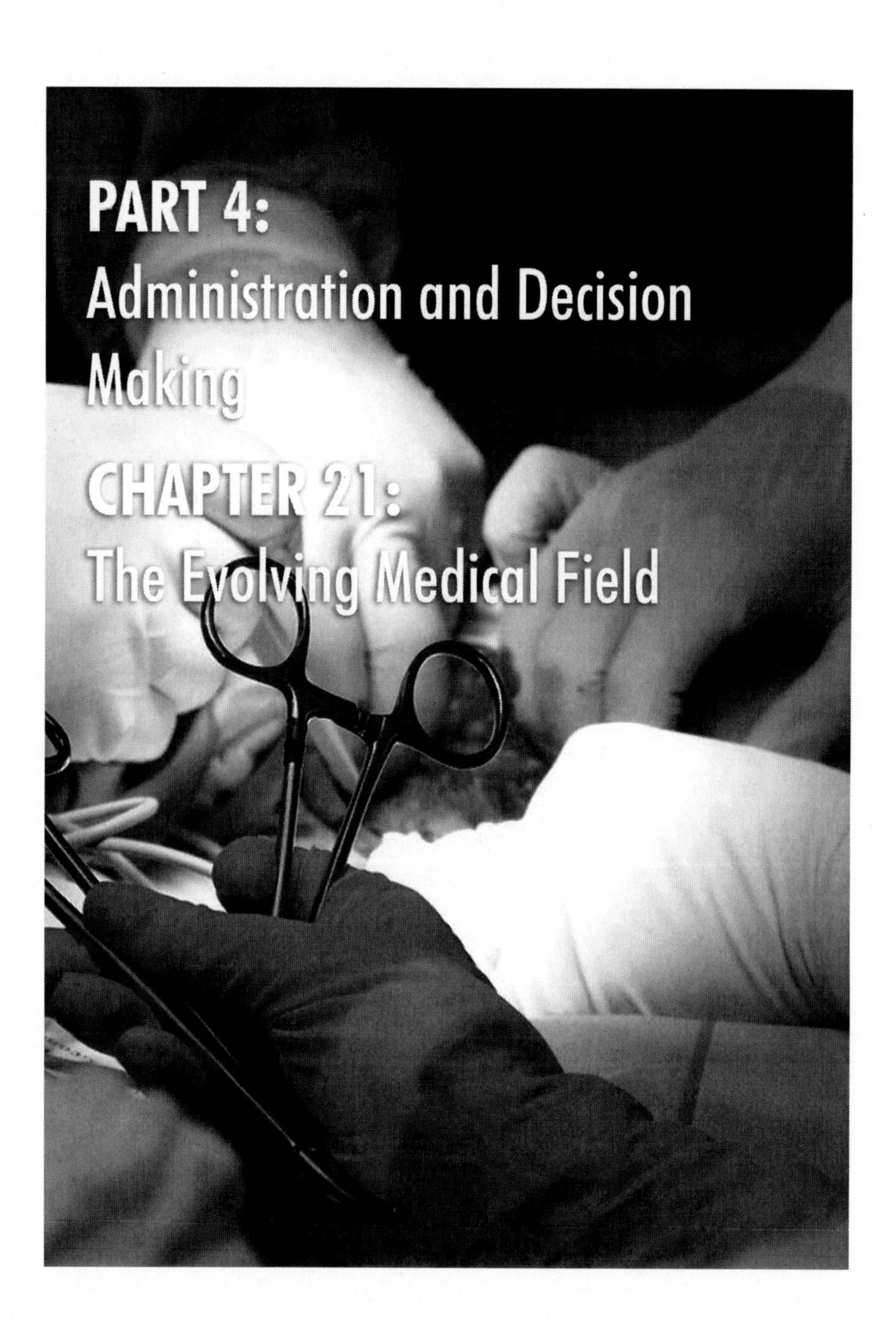

PART 4: Administration and Decision Making

CHAPTER 21: The Evolving Medical Field

The medical field has been evolving for hundreds of years and it is not slowing down with cancer research, genomic research, new medications targeting specific cancer cells, stem cell research, biomechanics, interventional techniques, robotic surgery, quality assurance development and so on.

In the 45 years of my training and practice, I have seen a plethora of developments, elements of progress and interventions in my field and in others. My secretary did not have a computer in the office when I started my practice. All these changes allowed me to learn, to grow, to provide increasingly state-of-the-art care, to become more critical of my work and my specialty, to become more efficient, accountable and even more patient-oriented with the development of family-centered care medicine.

Quality assurance is multifaceted and has become a large part of our practice. The medical field will not stop evolving and our learning curve after medical school and residency training is far from over. We need to be prepared to keep learning, to find time to keep up with changes and be adaptive and respectful.

Medical Training

Medical training has moved from its traditional approach, which was divided between preclinical and clinical studies. The former consisted of basic sciences such as anatomy, physiology, bio-chemistry, pharmacology, pathology, and the latter consisted of teaching various areas of clinical medicine such as internal medicine, pediatrics, psychiatry and surgery. However, medical programs have changed throughout the world.

In Canada, the Royal College has developed the Can MEDS Physician Competency Framework *(http://www.royalcollege.ca/portal/page/portal/rc/public),* which describes the knowledge, skills and abilities that physicians need to learn in order to provide patient care. It was developed in 1996 and incorporated in all post-graduate medical education programs in all of the 17 medical schools across

Canada in the early 2000's. Since 2009, it also included in all Family Medicine Programs in Canada and in other universities around the world.

The Can MEDS framework is based on 7 roles:

1. Medical Expert
2. Communicator
3. Collaborator
4. Manager
5. Health Advocate
6. Scholar
7. Professional

For each role there are key competencies to achieve and with each, there are enabling competencies to provide learning guidance. The Royal College is now developing the Can MEDS 2015 Competency Framework milestone guide which will align the existing framework with the next generation of competency-based medical education. It represents a new format of continuum learning. You should review the competency continuum diagram depicting the next generation of competency-based learning phases. (See website reference at the end of the chapter). Reading the Royal College 2015 Framework document will help you understand the new phase of learning, as this competency-based learning has a time factor since you will need to have learned certain competencies at each stage before moving to the next stage. This new learning process may have implications for your personal learning ability. Medical schools across Canada will move toward adapting the Can MEDS 2015 milestone guide over the next few years.

The evolving medical curriculum brings a new issue that medical students are facing and continue to face as early as the middle of the 3th year of medical school. By then, the trainees must choose a subspecialty to apply to in the match of their choice if they want a good chance to be chosen in the match and be funded. They may not have done their rotation in this specific subspecialty yet they have to apply for the match.

It may be a difficult and stressful experience for the trainee who does not or may not have an established idea of the subspecialty he/she is interested in. Matching applies to many other medical and surgical subspecialties, and one needs to recognize that matching is not only being accepted in the subspecialty of your choice, but also being accepted in the city/training program of your choice and being funded. The process can create a lot of anxiety early in medical school and the residency training as your career will depend on your choice. Starting in one discipline and finding that if you find it is not meeting your goals, you may apply to another subspecialty. Unfortunately, the process becomes complicated as all positions and matching has already been done and budget allocated to each program in relation to the number of trainees. It can be done, but it is not an easy or simple process.

I have talked about an academic career as a possible choice of medical practice when you complete your training. Trainees looking toward an academic/university appointment need to understand the large emphasis on research over patient care as the road to recognition, to university promotion and leadership positions, so one needs to carefully consider the difference between academic/university versus community/hospital-based career.

I personally believe that if you should pursue your training mindfully and along the way you will find what your interests are. If an academic aim is what you would like to pursue, opportunities will open up.

Quality Assurance

Medical care is more complex, new technologies offer improved diagnosis and treatment, patients are living longer managing their chronic medical conditions. There is an explosion of knowledge, new equipment, technology, research and medications all challenging physicians to remain current. This complexity has increased the expectations of the public and patients regarding the quality of care they can expect from their doctors and the health care system.

Over the years, hospitals and Canadian health authorities have established quality assurance programs to assess the delivery of health care services across Canada. Our health care services are paid for by the government (federal and provincial) so not only the public and patients but also health authorities have the right to ensure that good health care is what the public gets for their money. Our Canadian Health System is built on universality, a social responsibility and, in fact, is an obligation of our government.

Quality assurance encompasses credentialing, licensing, privileging, professional practice, morbidity and mortality review, quality of care assessment, electronic patient record information, informed consent and internet information, effective use of scarce resources, family-centred care, use of new technology and new treatments, accreditation of programs and hospitals, and more.

Quality assurance in health care is striving to reach excellent standards of care. It involves assessing the appropriateness, effectiveness and efficiency of medical and surgical tests and treatments. It is the measurement of health care improvement in all fields of medicine in a defined medical setting, be it a practice, a division, a department, a hospital. The concept includes the assessment or evaluation of the quality of care, identification of problems in the delivery of care, designing activities to overcome these deficiencies and follow-up monitoring to ensure appropriateness, effectiveness and efficiency. Since this book is not specifically about aspects of quality assurance, I will provide some thoughts on areas I have encountered in my practice and areas that are becoming important.

Credentialing, Licensing and Privileging

Upon completion of our training and successfully passing the examinations, the Royal College of Physicians and Surgeons of

Canada will provide our certification and credentials stating we have met the standards of our specialty requirements. The Canadian Royal College has multiple routes to certification to accommodate the various training pathways (Canadian, USA, or international students). The Royal College of Physicians and Surgeons of each province define licensing and privileging of doctors based on the competency standards and credentialing by the Royal College of Physicians and Surgeons of Canada, the Medical Council of Canada and the College of Family and Physicians of Canada. The health authorities and the Provincial Royal Colleges of Physicians and Surgeons have unique mandates and are regulated by different legislation but share the responsibility to assure the quality of services provided by medical professionals practicing in the health authorities.

As doctors, upon receiving our certificate of competency/specialty and credentials from the Canadian Royal College, we set out to find an employment location in which to practice in Canada, then we apply for licensing and privileging to the Provincial Royal College and the hospital we want to be affiliated with or practice at. The Provincial Royal Colleges have registration requirements including the credentials information from the Canadian Royal College before a physician can obtain a license (licensing) to practice in the specific province.

As a practitioner or a specialist, we really enjoy our practice with little review of our skills, knowledge and expertise. Quality Improvement is about improving the quality of our work through life-long learning and self-regulation. The Quality Assurance brings a joint responsibility of the profession through the provincial College, the RCPS, the Hospital Administrative Structures of Bylaws and Rules. Introduced in 2000, the Maintenance of Certification Program (MOC) was designed by the Royal College to support the lifelong learning needs of doctors. MOC is a framework to help doctors with their learning activities and a credit system to quantify these activities. This means our own quality improvement efforts and participation in quality assurance. In working to improve quality, our professionalism includes lifelong learning and self-regulation.

Check the Royal College of Physicians and Surgeons of Canada website to familiarize yourself with all the programs offered by the Royal College *(http:/www.royalcollege.ca/).*

Although the MOC is not a rigorous process to review doctor's skills and competence, it certainly provides help with continuous medical education. Ultimately, continuous medical education is left to each doctor to remain current and upgrade their skills through our personal ethics, accountability and responsibility. Our performance assessment is really done through processes such as a group practice, a division or a department in a hospital-based practice or an academic practice. Quality Assurance is the joint responsibility of the profession through provincial legislation, the provincial College, the RCPS, the HA structures of bylaw and rules.

Sometimes, unfortunately, our performance may be assessed through the media with medico-legal issues. National and international associations (which we can be a member of or a key leader, as there are several in all fields of medical specialties) provide excellent framework for data review, analysis and publications, which information we can use to assess our own performance and results.

As we get older and more experienced in our practice, we can provide support to our young colleagues starting their practice and in fact I believe thought should be put towards mentorship as a formal obligation. Doctors have monitored themselves through their respective specialty, organizations, their respective family practice group, literature review, attending meetings, implementation of database to track results, publications of one's own results, continuous medical education and quality assurance.

This multifaceted approach has been the backbone of quality assurance in our profession. It has not been a perfect process as we are all aware of conflicting results in the literature, of national and international medico-legal inquiries. This explosion of data and information has led to review of performance in relation to the number of treatments or surgeries doctors do in a period of time and what their results are.

This information is known among practitioners but also is slowly becoming slowly available for the patients and public. For instance, pancreatic surgery is a complex surgery, with careful indications and required surgical experience. All organ transplants demand an experienced team. Breast reconstruction surgery demands reconstructive plastic surgery techniques and skills. Very complex pediatric cardiac surgeries cannot be done in a small pediatric cardiac program. We can go on with hundreds of examples but the bottom line is that patients and the public will become more demanding, and we owe them provision of the best care and the best results.

The literature is overwhelmingly supportive of a critical mass/ number for doctors, be it a number of patients seen with a certain illness to maintain expertise, a number of surgical cases done per period of time to maintain skills and results, and so on.

Performance will be based on excellence, not just on good results. Canadian Health Authorities in association with the Royal College and hospitals will be more involved in the quality assurance of care delivery and credentialing and licensing. Privileging may well be matched or restricted to a more specific type of practice in your own field. It will not be left just to our profession anymore to "police" our practice and our specialty but to the Royal College and the Health Authorities to set stricter credentialing, licensing, and privileging standards, to grant specialty privileges based on scope of practice, to review licensing information periodically and perhaps restrict privileges based on competencies, skills and performance and/or changes in your scope of practice in relation to the development of new technology and treatment. Each province across Canada is working on developing assessment tools to review the core and non-core competencies of each specialty and how the ministry of health/Royal College/provincial licensing authority can anticipate changes over the coming years, and at the same time ensure that any specialist can train in what has become a non-core competency in his/ her specialty because of development of new techniques/technology/ treatment. Our future practice will be more scrutinized than ever before. We cannot avoid these changes and that is for the best as it will continue to push the envelope of care delivery for our patients.

The Ethic of Informed Consent

All of we doctors do some types of procedure where a consent is required. On one hand, we want to reassure the patient; on the other, we do not want to scare them off. The same confidence that allows us to do our job can also skew our perception of how well the patient and/or the family understands what is happening. Self-confident doctors like to get on with their work; others will somewhat play down the risks, complications and mortality of a procedure or may feel they provided the patient with proper information if they listed all risks and complications.

We are all caught with the feeling, rightly or wrongly, that less is more, and that too much information will be overwhelming for the patient. Many studies support the fact that most patients in a situation of dependence or worst emergency will retain just a limited amount of information and sometimes not even the most relevant information.

But increasingly in an era when patients are learning to research the information they need on line, patients are more ready for dialogue with their physicians. They come more informed with their questions and notebooks. Progressively, more patients are asking to record the conversation they have with you in your office. It never happened to me but with the prevalence of smartphone accessibility, it has become more current and it is the patient's privilege to understand the information provided to them so they can review it for themselves or with family members. Certainly in Canada, thankfully, it is not nearly the concern for potential medico-legal actions as in other countries such as the United States. Please refer to the article in the references "Healthcare in a land called PeoplePower: nothing about me without me."

A Surprising Question:

"I was on-call and I had done an emergency surgery during the night and got home at 3 a.m. in the morning. I stopped by the preoperative area to say a few words to the patient before his morning surgery. The patient asked me if I had a good night's sleep. My answer was: "not too bad" and I did not mention that I had spent most of the evening and night operating. I did not want to cancel the surgery and my colleague was not around."

There was no mention of sleep deprivation during my training 40 years ago and long weeks of more than 80 hours were common. I can sincerely say that I did not complain as I felt it was part of the training. And I have been on-call many hours, nights, and weekends during my practice.

Over the years, multiple studies have shown an increase in inappropriate decision-making, an increase in surgical complications following on-call nights and sleep deprivation. These studies have led to changes in the medical training curriculum with a decreased number of work hours per week, decreased number of calls per month, and in some subspecialties, the day after on-call duty is a day off. The Royal College of Physicians and surgeons of Canada has tried to balance the number of working hours, learning capability and sleep deprivation. This approach has been welcomed.

Conversely, there are several publications challenging this behavior that surgeons ought to disclose to their elective surgery patients when they have been up all night. They argued that patients have the right to choose whether to proceed or not with the surgery or re-schedule.

There are two camps: one, that we need to treat others as we would want to be treated; the other, that patients don't need to know. Opponents of disclosure argue that this would mark the beginning of a slippery slope. Those in favor of being honest with patients were mostly younger doctors. It is an indication that the culture of medicine is rapidly changing but one also cannot forget that the financial resources to implement such a policy of rescheduling patients and the realities of a hospital medical practice with long waitlists, staff scheduling and bed utilization are the other aspects of the day-to-day reality. (See chapter 22's section on sleep deprivation).

The new generation of doctors wants to be more open and honest with their patients. They recognize the pitfalls of the practices of the past. They take a more holistic approach to healing; they are more forthright with patients about what they know and they don't know. They believe in full disclosure and shared decision-making and are more accepting of patients who refuse treatment, such as a terminal/palliative patient. Many young doctors are able to speak to nontraditional therapies and other health modalities that most older doctors don't even know about. It is just part of the changes in medicine.

Our approach to treatment may be questioned. Our results may be challenged. With the availability of the Internet, many patients come well informed with many questions. Sometimes, they may have too many questions; sometimes they may request a different approach than what you suggest. In general, patients are quite understanding and reasonable.

Interestingly enough, rarely was I asked early in my practice what my results were in my experience with this type of operation, or how many cases I had done. Strangely enough, I was asked these questions later in my practice when I had more experience and confidence. It shows changes of attitude in patients who are now more demanding and more informed. Today, you will face these questions in your first year of practice.

Keeping records is a major part of your practice. Every hospital has an electronic medical record. Unfortunately, they differ from one hospital to another under the same health authority, but the importance is to keep records of patient information, to transfer and communicate relevant patient information amongst doctors so mistakes are decreased, effective and efficient care is provided and the patient benefits. One may think it is cumbersome to fill computer screens of patient information on the ward, in the OR, in medical records. But all computer programs are getting easier to understand

and it is becoming a regular part of our practice. I am sure younger physicians are well adapted to this approach. Remember, we used to write down this information, or dictate it and correct it. Now it is more instantaneous as we fill it in at any of the computer stations in the hospital, in the outpatient clinic, in our office. Ultimately, it saves time and is a good practice to remain timely and accurate. Even the best memory forgets sometimes.

As a physician working in a health care organization, we do not have choice over the EMR in use but a family practitioner or any specialist has the choice to implement an office patient record of their preference. There are multiple office software products that will allow you to have all your patients' demographic, to have a booking system, to enter daily notes when you see patients and to provide a comprehensive billing system. Each province has different software products available.

You can ask colleagues, search on the internet, evaluate products before choosing one, or you can always develop your own if you have time and computer technical skills. I would advise you to review the products available and buy one that suits your needs. Also, likely you will be able to buy a maintenance/update package with the software so your program is always updated with new features. Each examining room in your office/offices should have a computer terminal where you can pull the information needed for the patient you are seeing and you can enter your note after the consultation.

Transparency

Transparency is crucial in our health care system for effective, efficient delivery of health care and cost containment. Many agencies/organizations are reviewing credentialing, privileging, and new standards of accreditation.

A patient should be able to look up general information such as the number of surgeries per diagnostics, the number of complications, the mortality, the number of infections, the number of readmissions, the average length of stay per diagnosis. Patient should be able to enter their diagnosis and search such information. Most of this information is available in different formats at divisional and departmental level, at QA hospital committee, and perhaps at the government health authority in fractitionous reports. We certainly could do much better at regional and provincial level to provide a website for the public to access this information. The public is demanding this information and I believe it will become more available in the years to come. It should.

Accountability

The New England Journal of Medicine reported that as many as 25% of all patients are harmed by medical mistakes. What is most interesting and disturbing is that despite changes and numerous efforts, error rates have not come down significantly. The gray zone of when to treat is clouded by a medical culture that favors action over patience.

Hidden economic incentives behind treatment recommendations are not helping providing the best effective and efficient care. When it comes to overall doctoring, good listening skills are both a powerful diagnostic tool and have the power to heal, but we may be too busy and rushed to listen.

The Canadian health care system is vastly different than the US system. Hospital competition in the US is fierce and related to revenue, where a reputation can be tarnished by a low level of safety score as well as a poor performance in medical and surgical results. In contrast, Canada gives leeway to having information and results being much more transparent.

For instance, the use of a checklist before doing a procedure has been accepted in all hospitals across Canada. The compliance has varied but is improving, in some hospitals increasing from 65% to 95%. It is to the culture of the organization but studies are on the way to assess the impact of the policy on surgical errors.

Aviation polices the competencies of pilots and the screening older pilots with regular physical exams for vision loss and physical impairment that might diminish their ability to fly safely. The FAA removes older pilots who have vision problems, impaired judgment or diminished ability to communicate effectively and rapidly. Pilots are in charge of people's lives but so are doctors.

Many doctors work until they die. We doctors don't really like to retire. We love our job, and many times we do not know anything else and have few hobbies. Medicine comes to define who we are and becomes our self-worth. It can be so consuming that along the way we forget to develop other interests. Even if we do, nothing can replace the respect of our colleagues or patients. With each passing year of being a doctor, the respect seems to increase. It is a good recipe for not letting go. Many times we hear: "You are too young to retire"; "How can you retire with all the knowledge you have?"; "Why retire? You will be bored!"; "What do you do all day since you retired?" and "I don't know how you can do it. I certainly don't have enough money."

You will face many more questions and comments. Many older physicians are still in practice. In many hospitals, there are older physicians whose skills have deteriorated but who refuse to quit. This attitude is thankfully less present and the younger generation should be able to plan earlier in their practice. Like the old saying, quit while you're ahead, it applies also in medicine.

Not known to many of us, there are some hospitals in the United States where there is an effort to put in place a process to require that doctors over 70 years of age take exams for vision, memory, and judgment. For doctors who still perform surgical procedures, this would also include basic manual dexterity and encroaching tremors assessment. It is just the beginning of this approach but, in the near future, it is not implausible that older doctors will be required to have medical and physical tests as conditions of employment. This will come to Canada through the Royal College of Physicians and Surgeons of Canada and its respective provincial district. Re-licensing and privileging may require a medical and physical exam after 70 years of age or what ever age is determined by the College.

Family-Centered Care

Patient- and family-centered care is an approach to the planning, delivery and evaluation of health care that is grounded in mutually-beneficial partnerships among patients, families, and health care practitioners. It is founded on the understanding that the family plays a vital role in ensuring the health and well-being of patients of all ages. The ultimate goal of patient- and family-centered care is to create partnerships among health care practitioners, patients and families that will lead to the best outcomes and enhance the quality and safety of health care.

Family-centered care is based on four core concepts:

1. Dignity and respect:

    i) Health care practitioners listen to and honor patient and family perspectives and choices.
    ii) Patient and family knowledge, values, beliefs and cultural backgrounds are incorporated into the planning and delivery of care.
    iii) Health care practitioners communicate and share complete and unbiased information.

2. Information Sharing:

    i) i) Information with patients and families is shared in ways that are affirming and useful.
    ii) ii) Patients and families receive timely, complete and accurate information in order to effectively participate in care and decision-making.

3. Participation:

   Patients and families are encouraged and supported in participating in care and decision-making at the level they choose.

4. Collaboration:

   Patients, families, health care practitioners and hospital leaders collaborate in policy and program development, implementation and evaluation; in health care facility design; and in professional education, as well as in the delivery of care.

As an example of a center that appears to have a fully integrated family/patient centered care approach, Mayo Clinic's original principle of patient-centered care was set forth to provide the best care to every patient through integrated clinical practice. The culture has been established for years and is maintained by ensuring that secretarial administrators coordinate care. Instead of just filling another appointment at another date, doctors and their secretarial staff coordinate in that someone in every department is always available for walk-in appointment to ensure that a patient's visit is efficient and meets all his/her needs, and is even more important if the patient is from out-of-town.

Our culture of patient-centered care has been around for many years but we cannot in our current hospital system we realistically cannot deliver walk-in appointment on a daily basis. We would need a major reassessment of our resources, flexibility of clinic time, union agreement, doctors' and staff's cooperation. But we, as physicians, can make an effort to follow this culture in our daily practice, office, and hospital where we work, by simply doing our best to meet each patient's needs. Our culture defines the quality of the patient's experience.

"Mr. Jones is coming to see his family doctor for an upper respiratory tract infection. After a few questions and physical examination, the doctor feels Mr. Jones has a mild case of bronchitis and prescribes some medication. Before leaving, Mr. Jones mentions that he has problem with his eyes. The doctor explains he is too busy and his waiting room is full. "Can Mr. Jones talk to his secretary on the way out for another appointment."

"Mrs. Peters brings her 2-year-old son for an orthopedic assessment of his foot deformity. The orthopedic surgeon assesses her son, and orders some x-rays. She also mentions that her family doctor had heard a cardiac murmur at the last visit. The orthopedic surgeon explains it is not his domain of expertise and she can ask her family physician to make an appointment with a pediatric cardiologist."

A patient may have a good experience/visit with a physician but have to wait another month or two to see another specialist in the same hospital or, worse, the same office. A true patient-centered care culture would allow this patient to meet/see the specialists he/she needs in the same visit. It is the optimum patient-centered care culture that we should all strive for, and it starts with simple actions.

An integral part of a family-centered care program is to allow and encourage a designated family member to spend time with the patient, including spending the night. All organizations should strive to provide comfortable sleeping with pillows and blankets. Some doctors and nurses may feel it is a hindrance and nurses are well trained to care for patients.

Family members can be educated as to what warning signs to watch for and communicate with nursing staff. In addition, a designated family member can be invited to join the doctors and nurses for rounds in which they discuss the patient's plan of care. The family member should be encouraged to ask questions. When our ICU allowed parents to attend rounds or, more specifically, to remain at the bedside of their child during morning round, I personally thought it would be disruptive and slow us down with more questions. In fact, I was happily surprised that it was not disruptive; parents got more information and were very grateful.

Family presence is particularly helpful in reducing one of the major issues: falls. "Mr. Mitchell was admitted for chest pain at 85 years old. As the emergency room was very busy, his daughter could not stay all night with him. He tried to get up during the night and got over the bedrails, fell, and broke his neck."

Falls by weak and disoriented patients injure and even kill many people per year. A common sense policy to encourage family members to stay with their loved one could save lives.

Medical Errors and Checklist

The increased safety for patients provided by a checklist process has improved quality of care and decreased medical errors. The airline industry and the Toyota process have both been instrumental in developing this approach. It is an approach with less application in a family physician's practice, although I am sure the knowledge of it can be useful; the main benefit of a checklist process is in the complexity of medical and surgical care delivery. Medical and surgical mishaps are an important cause of mortality and morbidity.

Serious problems/complications will occur in 3-16% of patients undergoing a medical/surgical procedure. Half of the adverse events will occur during a procedure, specifically in the operating room around a procedure, thus the importance of focusing on teamwork, communication, adherence to good practice, and anticipation of events.

In 2009, the World Health Organization *(WHO: New Scientific Evidence Support WHO findings: Surgical safety checklist could save lives)* created a team to look at surgical mortality and morbidity in developed and developing countries. The team developed a surgical checklist and implemented it in eight hospitals around the world. Both mortality and morbidity decreased after the implementation of the checklist ($p<0.001$). The publication of that study and WHO guidelines identifying multiple recommended practices to ensure patients' safety lead to the use of this checklist and these guidelines in Canada and many other countries.

Over the last 5 years, compliance of using the surgical checklist has progressively increased from 65 to 90%. Clinicians do not find it difficult to use a checklist and in fact it is not as time-consuming as many argued it would be. Skepticism is disappearing in the face of quality improvement.

The effect on mortality and morbidity has not yet been published as of the writing of this book, however the checklist process has certainly increased communication, awareness, and improved our perception of teamwork and safety; thus, in a field with so many human errors and potential for human errors, having nurses and other staff using a checklist and other mechanisms, acts as a safety net to avoid mistakes and can make a huge difference. Creating a culture in which communication is encouraged can make a big difference in patients' outcomes.

Hospitals can be an intimidating place for health-care workers, including a young physician. The hospital has a strong hierarchy. The nurses, residents, and even young doctors taking and carrying out orders, implementing decisions about patient care and treatments, don't have the expertise of older, experienced personnel, including physicians, and may question their own knowledge, if they feel something is not right, communication may prevent a mistake and one should never be afraid to ask questions.

I would strongly recommend reading two books: "Human Error in Medicine" and "To Err is Human". These books should help residents in training and young physicians to understand/study/review the complexity of medical/surgical treatments, the interaction of multiple processes, human behavior in complex and stressful situations and the field of quality assurance improvement in patient care.

In Canada, we can debate the role of the government in controlling and managing our health care. Transparency of information and quality assurance is the crucial point. To make transparency effective, efficient governments, hospitals, and doctors must all play a critical role in making accurate reports and records about patient care delivery to help them with decision-making for the best possible care to be provided to the public.

REFERENCES

1. Lingard L., Regehr G., Orser B., et al.,
   Evaluation of a preoperative checklist and team briefing among surgeons, nurses and anesthesiologists to reduce failure in communication
   Arch. Surg. 2008; 143; 12-18
2. WHO: New Scientific Evidence Support WHO findings: Surgical safety checklist could save lives
   http://www.who.int/patientsafety/safesurgery/en/
3. Haynes A.B., Weiser T.G., Berry W.R., et al.,
   A surgical safety checklist to reduce morbidity and mortality in a global population
   N. Engl. J. Med. 2009; 360: 491-499
4. Haynes A.B., Weiser T.G., Berry W.R., et al.
   Changes in safety attitude and relationship to decrease postoperative morbidity and mortality following implementation of a checklist-based surgical safety intervention
   BMJ. Qual. Saf., 2011; 20: 102-107
5. Bogner M.S.,
   Human Error in Medicine
   Lawrence Enlbaum Associates Inc. Publishers
   New Jersey, USA, 1994
6. Kohn L.T., Corrigan J.M., Donaldson M.S.,
   To Err is Human, building a safer health system
   National Academy Press, Washington, USA, 2000
7. Makary Marty
   Unaccountable
   Bloomsburry Press, New York, USA, 2012
8. Binkmeyer J.D., et al.,
   Surgeon volume and operation mortality in the United States
   New England J. of Med., 349; 2003: 2117-2127

9. Nurok M., et al.,
   Sleep deprivation, elective surgical procedures and informed consent
   New England J. of Med., 363; 27; 2010: 2577-79
10. Cochrane D.D.,
   Investigating into medical imaging credentialing and quality
   http://www.health.gov.bc.ca/library/publications/year/2011/cochrane-phase1-report.pdf
11. Delbanco TI, Berwick DM, Boufford JI, Edgman-Levitan S, Ollenschläger G, Plamping D, Rockefeller RG
   Healthcare in a land called People Power: nothing about me without me.
   Health Expect. 2001 Sept;4(3):144-50.
12. Bilanow T.
   When Older Doctors Put Patients at Risk
   http://well.blogs.nytimes.com/2011/01/24/when-older-doctors-put-patients-at-risk/?_r=0
13. D. Kahneman
   Thinking Fast and Slow Anchor Canada Publishing, 2013

PART 5:
Family and Retirement

CHAPTER 22:
The Slippery Slope

Expectations and Reality

Most physicians entered the medical field believing that hard work and dedication would provide them with the knowledge and skills they need to provide the best care for their patients.

In the past, there was a direct career path and a person would just have to follow the path to meet their personal goals of having a great career. Changes in the health care system have happened in the last 30 years, resulting in an increased workload, more patient hours, less recognition for hard work, increased chance of litigation and, for many, decreased financial rewards.

Excessive cognitive demands caused by the need for quick processing of overwhelming amounts of information for long periods can add to a person's stress. Rapid evolution in the practice of medicine, increased patients demands, technology, growing bureaucracy, workload, conflicts, legal issues and level of health care changes: all have increased the complexity in our careers. In the near future, it will not be getting easier. There will be more pressure, greater demands and fewer people employed to do the work due to cost cutting that most organizations need to go through. Budgets are running very close to the minimum and need to be balanced despite demands on patient care, technology and personnel. The only aspect that one can count on is the ability to cope with change, to adapt to ever-changing policies and technology and to shift the practice to adapt. Because of all these changes, more work is done as a team and constant learning is required.

Therefore, as our careers evolve on a daily basis, there is a certain amount of uncertainty in our environments, in our practice and our future plans for retirement.

Contemporary medicine is intense and we all face a daunting uphill struggle against piles of paper, less time, increasingly demanding patients, complaints, managing our own practice and what appears to be doing more and more. It is, unfortunately, the direction of almost everyone's career. Some people say: "it goes with experience and age", but I am not sure it is true anymore. The reality of day-to-day clinical practice is different than the residency training we received and perhaps what we were told in school.

Physicians work in emotionally-charged situations associated with suffering, fear, failures and death, culminating in difficult interactions with patients, families and staff. Health professionals are held by society and also hold themselves to high standards of skills, knowledge and performance. It is believed they should always be at the peak of their ability to do whatever is needed and be emotionally available and compassionate. The rewards for this level of commitment are status, admiration, respect, proper equipment, space to do our jobs and financial compensation. All of these expectations are changing, as there has been a dramatic decrease of control and autonomy in patient-physician relationship, conflicts with peers, staff and administration. It has been abrogated in multiple ways, yet we continue to fulfill our side of the contract with a sense of confusion, disbelief and wondering where it will lead us. All of these factors contribute to increasing levels of anger, frustration, stress and burnout.

Burnout

Burnout is a syndrome and can be a combination of exhaustion, emotional and physical, which can lead to a low sense of accomplishment, decreased job performance, reduced job commitment and low career satisfaction. Burnout has many characteristics including: fatigue, exhaustion, inability to concentrate, depression, anxiety, insomnia, irritability and increased addictions. The most distinct characteristic is a loss of interest in work and family life.

Factors independently associated with burnout include younger age, having children, area of specialization, numbers of night on-call per week, numbers of work hours per week, and having compensation entirely based on billing.

1. A nationwide, multispecialty survey of more than 2000 physicians in 2011 found that almost 87% of respondents felt moderately to severely stressed and almost 63% admitted feeling more stressed and burnout now than they had felt 3 years ago. *("Mid-career burnout in generalist and specialist physicians", "Burnout and career satisfaction among American surgeons", "Results from CMA's 1998 physicians survey point to a dispirited profession", "Physicians Burnout")*

The trend from current studies still suggests the rate of emotional exhaustion, fatigue, tiredness, excessive amounts of work, lack of personal and family life balance is around 30-38%. It is supported by national samples of members of subspecialty societies. Almost 2/3 of Canadian physicians (62%) have a workload they consider too heavy and more than half (50%) felt their family and personnel life has suffered; 27% said they were on-call too often. Among rural physicians, 43% agreed that a lack of locums has affected their ability to take family and vacation time. Studies suggest that difficulty balancing personnel and professional life, administrative tasks, lack of autonomy and patient volume are the greatest sources of stress.

As stated earlier, major factors that act as stressors in our profession are information overload by the rapid advance in information technology, by the information explosion in the literature, on the internet, which is accessible not only to us, as physicians, but to our patients. The rapid advancement of technology has necessitated the development of updated knowledge and skills in diagnostic and therapeutic procedures. Our workload has expanded and at the same time we are working with shrinking resources and budgets. Other factors we are dealing with are increased consumer demand and expectations from patients. It can also be a deficiency or lack of educational opportunity, free time ability to ventilate, poor institutional support, inter-personnel pressure caused by managing our office, surrounding hospital staff, disruptive meetings and, surprisingly, sometimes having to handle a strong emotional reaction following a patient's death. It happens to all of us. As the stress builds up burnout can be just around the corner.

Sleep Deprivation

Fatigue from sleep deprivation may be due to the loss of one night of sleep, chronic insufficient sleep (over several nights), repeated interruptions of sleep which may be attributable to working long shifts, long work weeks, a sleep disorder or personal circumstances. It is a challenge for medical systems to ensure that physicians do not have sleep deprivation, given competing needs for the continuity of care, 24/7 coverage of clinical care and emergencies.

To implement any policies to support physicians, institutions will need to absorb financial and administrative consequences of cancelling and rescheduling elective surgeries in a timely manner.

In keeping with the ethical and legal standards of informed consent, patients awaiting a scheduled elective surgery should be informed about possible impairments induced by a lack of sleep and the increased risk of complications. It is a fact that sleep deprivation can interrupt one's ability to recognize or assess accurately the risks posed by an intervention or the risks occurring during an intervention. Therefore the patient should be given the choice of proceeding with surgery, rescheduling surgery or to proceed with a different surgeon.

Even when physicians understand the potential increased risks of complications performing surgery with sleep deprivation (and the literature does not have a published cause-to-effect complication rate), there are multiple problems to rescheduling a surgery:

1. The patient is ready for surgery, has made arrangements with family members, work, travel, and any of these arrangements may be difficult to reschedule.
2. The surgeon may have several patients that day and when one surgery is cancelled, his busy practice may find rebooking difficult or time may be limited, pattern of practice may be

difficult to change with limited surgical staff, coverage time and finally loss of income.

3. Rescheduling surgery at the last minute is challenging for the operating room with decreasing resources and operative time.
4. For the institution, the waiting list may increase, there will be challenges in the booking system, and there may be discontent and complaints from the public.
5. It may and will lead to inefficiency, ineffectiveness and increased costs to the health care system.

Let's face it, we have all been in this situation more than once and we will be there again. It is unfortunately the current state of affairs, but we do not have to be complacent. We can continue to make the effort to improve patient safety at surgery.

Stress and Addictions

Besides the toll stress can take on your personal life. Burnout and fatigue are known to adversely affect performance efficiency and promote behaviours that have a negative impact on staff relationships and patient care. Too often, we hear from our peers (such as nurses, colleagues and allied health professionals) that these high levels of stress are part of the territory of working in the medical environment and are not symptoms of stress and burnout: cynicism, arrogance, anger, lack of control, depression, workaholism and disinterest. When these symptoms start to appear, consider these indicators as the time to take steps to address stress in your life and workplace.

Physicians deal with other people's problems all day, but they are least likely to recognize or raise their own difficulties. Approximately one-third of physicians do not have a family doctor. If we do not take proper care of our own physical health, how can we even start to take care of our emotional health? Many deny their own emotional needs as a survival mechanism. Many physicians follow the implicit road of protecting themselves by not allowing themselves to feel too much emotion, sympathy or sadness. It is not a good option. So stress from these unfortunate, appropriate, required changes in the environment of health care can insinuatingly invade our life: not just work life, but all aspects of our life.

The first step is for the physician is to recognize that a problem exists. Physicians work under stress most of the time. They are so heavily involved in their work and preoccupied with patient care that the thought of being under stress rarely registers. In effect, their focus on work and patient care reduces their sensitivity to and awareness of the impact on others that their actions and behaviors have. Physicians may recognize in themselves the more obvious physical symptoms of stress such as chest pain, palpitations, headaches, muscle pain, panic/ anxiety attacks, and gastrointestinal distress.

Frequently, however, they do not recognize the more subtle symptoms such as irritability, mood swings, anger, loss of focus, sleep disturbance, unhappiness at home and an overall sense of not being happy. Once physicians recognize they are under stress, very often their usual position is that they can handle it themselves. They have lived with stress all their lives and they can manage it just fine. They are reluctant to admit and share their inner emotional stress for fear of being seen as being weak, to have their performance criticized by the head of the department, or be unable to make decisions and provide proper patient care.

Although physicians' denial and reluctance are potential obstacles, there are effective ways to address, diminish and/or overcome these barriers, even if many physicians note that finding the time and the money to do something to relieve stress is a challenge. Before requiring treatment and/or rehabilitation for addictions, there are easy strategies for managing stress that include getting enough sleep, eating a proper diet, not snacking every night, watching TV or exercising and building some down time for relaxation and even daily meditation. Because time pressures and demands often consume the day, physicians must build in personal time as a priority. Do not over extend yourself by accepting more and more projects. It is quite acceptable to refuse added work, whether it is administrative, research, or patient work.

I personally went to the gym at 6.45a.m., 5 days a week. Everybody at the hospital knew my schedule and knew I would not attend a 7a.m. meeting. It did work for me. Other small changes can include choosing to sit down for breakfast, taking time for a healthy snack (even if it is in your office), reading non-medical journals or magazines, going for a 10-minute walk at night or going to see a movie. On a larger scale, plan a holiday you will enjoy. Not everyone likes outdoor activities and packed vacations but you can plan a holiday with your family that you will look forward to and enjoy. Just to put your feet up feels amazing.

Making time for family and friends is very important because they are your support system, they are who you love and some informal advice from your loved ones is also another way to reduce stress. We, as professionals, also have the opportunity for continuing education, research, and writing, collaborating with colleagues, mentoring trainees and going on a professional sabbatical. The important thing is to recognize to put these elements into our schedule so they represent pleasure and something to look forward to. There are many more examples and ideas, but all these activities recognize the value of establishing a work-life balance.

We tend to play down our own organization because of budget restraints, staff difficulty and regressive culture, but in reality our own organization has a responsibility to provide us, physicians, nurses and health care professionals, by providing a healthy environment and proper resources and equipment. Our legitimate complaints about issues affecting our practice should be addressed by the administration and not always the other way around. Organizations should be more helpful in reducing the stress of their staff by providing more ancillary and staff support, on-site exercise facility or daycare, workshops and education on managing and coping with stress and mentoring resources. These are all fairly simple strategies that our organization can provide to help cope with a stressful profession.

In most cases, early intervention and small changes through awareness are sufficient. Allowing physicians an opportunity to discuss practice and personal issues may be all what is needed. Occasionally more comprehensive stress management and substance abuse programs may be in order, such as anger management, conflict management, psychological and emotional management and ultimately a rehabilitation program. Early intervention through friends, family, colleagues or a more formal physicians' support program is preferred.

Substance abuse continues to be present in all sectors of society, so it is not surprising that it also occurs in the medical profession. Studies conducted among samples of physicians suggest that rates of substance use (alcohol, opioids, amphetamines and others) are between 10 and 15%, similar to the general population and a similar population of professionals. Alcohol was still the most commonly used substance (8 to 10%) of physicians. Of interest, except for a higher tendency to smoke, surgeons appear to have lower rates of substance abuse and are less likely to have a dependency on drugs during their lifetime.

The rates of opioids (opiates and benzodiazepines) dependence and abuse are higher among physicians, estimated between 1-2%; it is greater among anesthesiologists, psychiatrists, emergency medicine physicians and family practitioners. Despite changes in methods of reporting, of distribution, of accountability of opioids and mandatory education, it is unclear how effective these methods have been. The use of drugs such as cocaine, amphetamines and heroin is less than 1%. Prescription drugs may be higher due to their easier availability to prescribe and buy.

It is not easy to identify a physician suffering from alcohol/ substance abuse because the clues/symptoms can be very subtle and physicians are good at hiding things.

"Shelley was a busy health care professional, manager of several hospital programs and very well-respected for 20 years. She denied having an alcohol problem to colleagues until one afternoon after work she fell in the parking lot and was brought to the emergency ward. She was found to be in renal failure, liver failure and encephalopathy."

Addictive behavior by a colleague often goes undetected. Causes for delay in diagnosis include the physician's inability to accept having a substance abuse problem, the physician believing he/she may be able to control the addiction because of their medical knowledge, fear of disclosure of an addictive disorder for fear of losing prestige, fear of losing respect or losing his/her license and livelihood. Doctors may feel uncomfortable being a patient and may interpret their needs as a weakness. Also, a physician's family member or colleague will often participate in a conspiracy of silence in an effort to protect the physician and hoping that he/she will address the problem on their own.

The physician wants to preserve their work performance above all other aspects of their life and by the time a physician's addictive illness becomes apparent in the workplace, the rest of his/her social, family and personal life is in shambles. Sometimes a decision to help a colleague must be made with inconclusive evidence and confrontation, understanding it is what is best for the involved physician.

After a diagnosis of addiction has been established, treatment should be initiated in a program that specializes in the care of addicted physicians, as they offer specific modalities targeted to physicians' group therapy helping addicts through denial, the threat to their career, and their family life. Physicians have remarkable abstinence rates after completing a rehabilitation program compared to the general population: between 74% to 90% success rate. When a physician returns to work after addiction treatment, the hospital administration/department/colleagues generally can impose restrictions on work, type of work and amount of work, which should not be viewed as punitive, but as help to a successful recovery and retaining the respect of the physician.

Medical School Curriculum

Medical school and residency training are demanding programs but most students and residents will complete their training without considerable difficulties. However, some *will* experience difficulties. Issues that might affect a student's functioning include personal issues such as health, family, finances, culture and training-related issues such as adjustment, conditions of learning, workload, ethical conflicts, inadequate support and so on. These issues might affect training and, like stress in your practice, become apparent quite late in training.

The Accreditation Council for Graduate Medical Education (ACGME) in the US and the Royal College of Physicians and Surgeons of Canada accredit more than 8,000 residency programs in over 100 subspecialties. Awareness of sleep deprivation in residents-in-training has been made a priority at the national level: they recognize that shorter hours will make residents more alert and better able to learn. Over the last 15 years, the number of hours on-call, of working hours per week, of days off post-call, have all been addressed and multiple changes have been made. But these changes are also result in less exposure to patients, the course of their disease, patient care and even decision-making. Unfortunately, there are no scientific answers to resolve the tension between the optimal number of hours for education and hours for rest. Research and studies have not been able to establish an exact number of hours per week below, which residents may safely and effectively learn and participate in patient care. It is a difficult balancing act.

These changes in the residency training program include the hours per week worked, the on-call hours, the day-off post call and more time for sleep contribute to increasing residents' satisfaction and hopefully help learning and safety. But these changes have also created another problem. The residents are vital to the provision of patient care in a system that faces workforce shortages. A major challenge faces program directors and hospital administration in the difficulty of providing continuity of patient care caused by restrictions on residents' work hours. Efforts to reduce duty hours must also address the critical issue of providing clinical services with a reduced number of resident hours. Although most residency programs model resident hour reductions, many have replaced residents with other practitioners (mainly hospitalists). This approach is costly and may be complicated by a shortage in health care workers. At the same time, the health care system deserves appropriate patient care coverage by a combination of residents and medical staff. In the end, there has to be a balance between the residents' working environment and the high quality of care patients deserve.

Too often, teachers will notice a student having difficulties and will either avoid the problem or fail to share their concerns with the student early in the course of the training, to the detriment of the student who is often unaware of the concerns about their performance. Despite the numerous frameworks available that outline the evaluation and management of students facing challenges, most are vague, without specific tools to help facilitate the management of students facing learning disabilities.

Certainly for most of us, we do not have the proper training to assess these types of challenges, which the residents may be experiencing. Some physicians may not feel that it is part of their job to be looking for challenges with residents. At the same time, we cannot wait until a resident has failed to raise concerns. Many educators are starting to believe that we should adopt a more open teaching and learning approach. If it is inappropriate for a faculty member to develop a counseling and personal relationship with a resident, the faculty member should recognize the learning issues of the resident and then, possibly, refer the resident to the appropriate professional support.

It is crucial to obtain information about the challenging issues, previous rotations, educational experiences, study habits, familial issues and substance abuse/addictions if any exist. Physicians should document the specific issue with questions about which difficulties the residents have experienced, when the difficulties started, why they occur, and how severe the challenges are. Having been a program director myself, I strongly recommend that any meeting with a student is done with a colleague as a witness, that the student has an understanding of the goals of the meeting and that the student is told that notes will be taken and that he/she will sign the document with notes after corrections. Since students who are in difficulty are often not aware of their own challenges, discussing the problems identified by the teacher/mentor provides a good opportunity to appraise and promote self-assessment skills in the students.

However, as the patient-centered care model improves patient care, the adoption of a student-centered model under the CANMeds of the Royal College of Canada will improve teaching and learning.

Living the Good Life

In these times of rapid technological changes, people often feel out of control, powerless to stop change and/or navigate through their way through it. Stress is epidemic in our society. But aspects that you can control are your attitude and the health of your body. You can decide if you are going to be a positive person and you can decide if you are going to be fit and eat healthily. A commitment to physical activity brings balance to your mind, the intellect and the body. Knowing something is good for you is not enough to make you do it. If you just relax and listen to your body, it will tell you what it needs. Exercise can be a form of meditation by turning your attention away from work and from problems, bringing relaxation and reducing stress. For example, there are 168 hours in a week. If you exercise consistently 2 hours per week your body will feel better. You will increase your life expectancy by 2 years or more.

Work has changed to become less physically challenging and at the same time more mentally stressful. The images of the perfect body seen in the media and marketing are not real. Magazines and TV make money promoting fitness perfection but this high level of fitness is not realistic. Your exercise regime is not about perfection but about feeling good mentally and physically. Exercise will help you regain control and decrease stress in your life.

It is more important to control your eating habits than be on a diet. Cheese and crackers every night before dinner is not a good habit. Eating a box of cookies while watching TV is also not good for you. A few pounds here and there will catch up with you and your work. You will feel better by eating the proper number of calories for your lifestyle and body type. For instance, a 1200 calories diet may not be enough, even if you want to lose weight; with this amount of calorie intake you will lose muscle. Any weight control program that includes an extremely low calorie diet every day and goes to certain extremes with specific foods should be avoided. Proper weight control is a combination of exercise along with a balanced, healthy diet. This is the best, long lasting approach. Being under stress and overweight does not allow for "your best performance". The brain has a greater capability to deal with the demands of work and tasks if you exercise and are careful with your diet. Other aspects of your life will become easier too.

A private trainer may sound expensive, difficult to commit to or you may think you can just exercise sufficiently by yourself. But it is a small price to pay for a substantial investment: your health. He/she will help you to achieve your goals faster than on your own; you will receive an individualized program for your body type and level of fitness; you will maximize your training; you will be exposed to a variety of training exercises; your trainer can also protect you against injury, you will be stimulated and it is more fun working out with another person.

A good life is being in a state of balance. Balance is the ability to cope and deal with your work and life issues in a positive manner. Exercise, diet, and fitness will lead to balance in your life, where you are not always consumed by your work and where family life can be an important part of the work/life balance.

It is not necessary for us to sacrifice our own health, family and community in order to care for others. A preventive educational approach is now in place in most medical schools. Heightened awareness of stress management and substance abuse through prevention programs, medical community lectures and conferences are available and should be encouraged. We cannot appropriately relieve the suffering of others if we are suffering. Ultimately, individual physicians will personally benefit from taking better care of themselves and health care organizations will gain in patient care delivery, reduced absenteeism, job satisfaction, recruitment and retention by supporting the physicians.

Spirituality and Wellness

Spirituality is unique to each individual. Your "spirit" usually refers to the deepest part of you, the part that lets you make meaning of your world. Your spirit provides you with the revealing sense of who you are, why you are here and what your purpose for living is. It is that innermost part of you that allows you to gain strength and hope.

The notion that one can be spiritual but not religious is like saying you have blood in your body, but don't care much for the skeleton. Similarly, claiming you need to go to a place of worship to be spiritual is like saying you need to shop at Macy's to wear clothes. Both ideas are bunk.

Spiritual wellness may not be something that you think much of, yet its impact on your life is unavoidable. The basis of spirituality is discovering a sense of meaningfulness in your life and coming to know that you have a purpose to fulfill.

Many wellness behaviors can benefit your spiritual health. Such behaviors include feeling connected with others, feeling part of a community, volunteering, having an optimistic attitude, contributing to society and self-love/care.

Practicing Spirituality for Personal Growth:

You can use spiritual practices to strengthen your body, mind, and spirit. All three are connected. For example, your practices to strengthen your body also support the increased flows of energy that come from your mental and spirit-based practices.

- Practices for your body: These practices include yoga postures, breathing exercises, taking time for yourself and energy balancing.

- Practices for your mind: These practices include being quiet for a few minutes, meditation, study of spiritual teachings, readings.
- Practices for your spirit: These practices can include exploration of one's conscience, explorative readings, or even prayers and attending places of communion. Here, the truism is simple; to each their own.

Strengthening your Willpower:

Spiritual disciplines strengthen your willpower. For example, you may decide to meditate for a half-hour every morning and self-control is like a muscle that you strengthen through meeting challenges with your willpower. Gaining more self-control in one area also gives you the strength to free yourself from other harmful habits.

You can stay focused in your commitment to meditate for the full half-hour. This is just like exercising your muscles or your mind. Even if you sometimes give in and lose the battle, whatever efforts you have put into confronting that desire or habit or addiction still bring you greater strength the next time. Every time you conquer the pull of your senses, you become more in charge.

You can apply this same technique to other willpower situations, such as quitting smoking or maintaining an exercise program. The key is to make a decision and honor it, keeping your word to yourself.

Practicing being Happy:

You can develop the habit of being happy just by practicing acting happy. Your inner being and outer expressions reflect one another. If you feel happy inside, it shows up in everything you do. In the same way, expressing happiness outwardly also affects your state of mind, creating greater happiness inside yourself.

During good times, allow yourself to smile; during difficult times, still do your best to smile. It doesn't even have to be a big, outer smile. Just think, "smile." Think "contentment." Feel what it would feel like to have a spontaneous, sincere smile playing on your lips. It does not mean that you can disregard your grief or anger or walk around with a fake smile. It is will power.

Meditation:

With the hectic pace and demands of modern life, many people feel stressed and over-worked. It often feels like there is just not enough time in the day to get everything done. Our stress and tiredness make us unhappy, impatient and frustrated. It can even affect our health. We are often so busy that we feel there is no time to stop and meditate. But meditation actually gives you more time by making your mind calmer and more focused. A simple ten or fifteen minute breathing meditation as explained below can help you to overcome your stress and find some inner peace and balance. Meditation can also help us to understand our own mind. We can learn how to transform our mind from negative to positive, from disturbed to peaceful, from unhappy to happy. Overcoming negative thoughts and cultivating constructive thoughts is the purpose of the transforming meditations found in the Buddhist tradition. This is a profound spiritual practice you can enjoy throughout the day, not just while seated in meditation.

REFERENCES

1. Burnout and self-reported patient care in an internal medicine residency program
   Tait D. Shanafelt et al,
   Ann Intern Med; 2002; 136: 358-367
2. Sleep deprivation, elective surgical procedures, and informed consent
   Michael Nurok et al,
   N Engl Med;2010;363:27;2577-2579

3. Sleep loss and performance in residents and non physicians: a meta-analytic examination
   I. Philibert,
   Sleep; 2005; 28(11), 1392-1402
4. Mid-career burnout in generalist and specialist physicians
   A. Spickard, S.G. Gabbe, J.F. Christensen,
   Jour. American Med. Ass., 2002; 288, 1447-1450
5. Burnout and career satisfaction among American surgeons
   Tait D. Shanafelt et al,
   Ann Surg; 2009; 250:3; 107-114
6. Results from CMA's huge 1998 physicians survey point to a dispirited profession
   Patrick Sullivan et al,
   CMAJ; 1998;159;525-528
7. Physicians Burnout
   Linda Gundersen
   Ann Internal Med; 2001;135:2;145-148
8. Burnout in physicians: a case for peer support
   S.M. Bruce, H.M. Conaglen, J.V.Conaglen,
   Int. Med. Journal; 2005; 35, 272-278
9. New requirements for resident duty hours
   Ingrid Philibert et al
   JAMA 2002; 288:9; 1112-1114
10. The well-being of physicians
    T.D. Shanafelt, J.A. Sloan, T.M. Haberman
    American Jour. of Med., 2003; 114 (6), 513-519
11. Physician Wellness: a missing quality indicator
    Jean E. Wallace et al
    Lancet; 2009; 374; 1714-1721
12. Lifetime psychiatric and substance use disorders among impaired physicians in a Physician Health Program: comparison to a general treatment population
    Linda B. Cottler et al
    J Addict Med; 2013; 7; 2: 108-112

13. Substance abuse among physicians: a survey of academic anesthesiology program
John V. Broth et al
Anesth Analg; 2002; 95: 1024-30
14. Physician substance abuse and recovery
David R. Gastfriend
JAMA; 2005; 293: 12; 1513-1515
15. Substance abuse and dependence in physicians: detection and treatment
Patrick Asubonteng Rivers et al
Health Manpower Management; 1998; 24: 5; 183-7
16. Chemical dependency and the physician
Keith H. Berge et al
Mayo Clinic Proc; 2009; 84: 7; 625-631
17. Suicide rates among physicians: a quantitative and gender assessment (meta-analysis)
E.S. Schernhammer, G.A. Colditz,
American Jour. Psychiatry; 2004; 161(12), 2295-2302
18. E. Tolle
The Power of Now: a guide to spiritual enlightenment New World, 1999
19. E. Goldstein
The Now Effect
Atria Books, NY, 2012
20. W. M. Sotile, M. O. Sotile,
The Medical Marriage, Sustaining Healthy Relationships for Physicians and their Families
AMA Press, 2000

PART 5:
Family and Retirement

CHAPTER 23:
Is Retirement Really Possible?

There is no real magic around planning your retirement. You can start planning the month you establish your practice or you can wait until much later in life. Although few people think about retirement at the start of their career, a few steps can be taken early on that will help you a great deal.

Savings Strategies

The most important point is to ensure that you put away the maximum contribution you are allowed each year toward your Registered Retirement Savings Plan (RRSP). Your accountant will be able to tell you what you can contribute to your RRSP's every year after they do an analysis of your current tax liabilities. RRSP savings are tax-deductible from your personal income tax and the same applies to your spouse's personal income tax. It works similar to a pension plan, where you withdraw your money after the age of 65, but will be taxed in full when you use or draw money from your RRSP.

As you will earn an average of $250,000 to $500,000 annually, the minimum RRSP contribution will be approximately $15,000. The amount invested per year for a minimum of 30 years at a conservative 5% annual return, has a potential of accumulating $1.4 million over 30 years. This simple step is easy to follow every year and will greatly benefit your retirement. Plan every year to have that amount available in your savings in December, so you have the funds to transfer to your RRSPs. If you do not have the money available and have to borrow money, then this defeats the purpose as you may end up borrowing money at an additional expense.

You should establish some goals and strategies. You will want to identify the goals you want to reach such as; when you want to retire, have your house mortgage paid off and perhaps your vacation home mortgage paid in full. It is a good idea to have an idea of how much money your portfolio should contain, realistic or not, to meet your goals and a plan for your children's education (perhaps establishing a college fund). At 30-35 years of age, you are thinking about establishing your practice, having children, getting a house, paying student loans and the idea of retirement planning may be low on your list of priorities, but at the very least, the RRSP contribution is a must.

Retirement Age

This may not be a concern of yours as you are establishing your practice, but as you get closer to 55 to 60 years old, you will be thinking about retirement more and more and this planning will pay off in the long run. You will feel more at ease and have a happier retirement the earlier you plan your finances for your senior years. Most doctors, of course, retain their skills and sharpness of mind well into their 70's, but physicians are not immune to illness and aging. Diminished competence is not only attributable to age but also to depression, substance abuse and a failure to maintain proficiency through continuing medical education. Physicians do have to meet minimal requirements to continue to practice.

Many physicians think they won't retire, but most will retire between 65 and 70 years of age. One reason why so many doctors do not want to retire is that medicine becomes their whole life. Doctors can define themselves through their medicine and work. With experience and years of work, we get more respected by colleagues and patients. It defines our life and certainly provides acknowledgement of our expertise and skills.

Patient advocates and some experts are calling for regular cognitive and physical screening once doctors reach 70 years old. These tests will measure physicians competency. This process will be similar to the airline industry, where pilots go through an annual assessment program, although in the airline industry they do this at much earlier ages. Medical professionals are supposed to report unsafe colleagues but some doctors are reluctant to confront their fellow colleagues or their seniors. It may take several more years to address this situation but some hospitals are starting to address aging physicians and implementing some forms of ongoing cognitive and physical exams. Whatever forms of assessment may be coming in the future, you do not want to be the one everybody is talking about, but being unable to quit.

You may not be sure of your retirement date as it may change for different reasons (for example a situation like the 2008 financial crash may affect your retirement portfolio and the planned age which you had thought you may retire), but the main goal is to prepare for your retirement by managing your financial situation carefully, so that you are not forced to work at 65 and if you do, it is for enjoyment, pleasure and fulfillment, which is much less stressful than working from a place of financial need.

Slowly reducing your practice might be a great solution for your retirement plan. Because medicine is a complex profession, ever changing and developing, knowledge and skills are better maintained with practice. You may have the possibility to decrease your amount of weekly hours if you are in a group practice or a partnership. Many doctors I know, specialists or family physicians, have decreased their workload to 80%, 75% or 60% of their regular practice after negotiation with their colleagues. Your colleagues may agree to your request or get extra help by bringing in a part-time colleague. You may work full time but stop being on-call. Options are available and the larger the group practice, the easier it is to exercise them.

With expertise, communication with other colleagues around the world, presentations, publications, and contacts, we have established a career which can span over 30 years, we have endless possibilities to work part-time, such as teaching, consulting, research, surgical assistance, legal advice, writing and volunteering.

Assets

I always say, "It is not how much money you have but how much money you spend". You must evaluate your financial situation from time to time. It is quite normal that as we get older, our expenses and obligations increase: life-style, cars, houses, trips, education and tennis/golf memberships, etc. – the list is long. The important question is: do you want to maintain the same lifestyle or do you plan to diminish it when you retire? I know many doctors who have told me "I worked hard for that lifestyle" and I agree. Then you will need the cash flow to maintain it.

The ideal situation is to pair your expenses with your assets-generating income. It would therefore allow you to live on your investment return and not touch the capital. In that calculation, you need to factor in the cost of annual inflation, as an investment that does not keep pace with the inflation is a negative investment and over 30 years of retirement, your capital will diminish. In order words, if your investment generates 5% return, and you use the total amount, you are using 2% inflation rate from your capital. To really use the 5% return money for your living expenses, the return on your investment needs to be 7%, 5% for your expenses and 2% for the inflation and your capital remains untouched. It is a concept that many of us, (including myself in previous years), did not grasp. Every passing year, our nest eggs can be eroded by inflation.

If we need to talk about how much money one should have, let's establish a principle that most financial advisers/books suggest. You should have enough income producing assets to be able to live off an amount equal to 5% of the total value of your portfolio per year. The days of 10 and 12% return per year on investment are probably over and for the years to come. There is still too much uncertainty in the world economy. So, 5% return on $2M asset is $100,000 per year. Is this enough to maintain your lifestyle? If not, you will have to decrease your expenses, or plan to save more money.

Several facts might influence your monthly income in retirement. Your house is too big for two, represents a lot of money and you are planning to downsize it and invest some of the money. You may plan to start to rent your vacation home or perhaps the basement in your home to generate extra income. Additionally, you may start receiving your university pension plan at the age of 65 years (if you have a university appointment or track tenure). At this time, your spouse or both of you may carry on with part-time work. And many physicians will state that with downsizing the house, closing the office and the decrease in expenses, i.e. no insurance liability, no life insurance, less gas, perhaps only one car, less clothes, less general expenses. All these facts need to be taken into account when planning how much money you will you require for your "new" retirement lifestyle.

Art, collectibles, antiques, even jewelry are an asset but not an income-producing asset. I am sure (most) physicians do not purchase Picasso or Rembrandt paintings, so the art we buy is entirely for our pleasure and will not produce a sizeable amount of return if sold.

Finally, there are real life situations that cannot be anticipated and may greatly affect our life: death of a child, of a spouse, a divorce, a bad investment. No one can really plan for such a traumatic and painful event.

Useful Advices

I would recommend that as much as possible, one must plan to retire debt-free or nearly debt-free. I realize that nowadays, it is becoming more and more difficult. But the less you owe, the better it is. If anything goes wrong, there is no income revenue and it would be ill advised to use the capital early in retirement. Regenerating the capital through future return usually falls short. You should ideally own your house and any other significant purchases such as a vacation home.

Retire with a new car you have partially paid through your practice so you can postpone this expense for several years. If you plan to stay in the same house, when you retire, review the possibility of major repairs or renovations before you retire. Rental properties such as condominiums or apartments are a good investment. Unless, the mortgages are paid and the rental income is substantial, you may want to consider selling it to obtain a good return on your investment. But hopefully, your properties have increased in value over the years and the investment will have paid off.

The other principal obligation is your children's education.

This retirement issue might be more challenging to achieve as you might have remarried or married late in life, and you still have teenagers in college. Hopefully, you have planned for retirement and working a few more years is not necessarily the solution you are forced to choose.

Credit Cards

We live with credit cards and we tend to forget how much money it may represent on a monthly basis as we pay some amount down every month but probably not the full amount. For some doctors, it is common to have 2 credit cards, one personal credit card and one for business. You may also have two in the name of your spouse.

As we get older and establish a good credit history, and become more established, the banks love to increase our limit so we can spend more. Frequently, the total amount on our several credit cards, adds up to more than expected, and can easily be over several thousand of dollars. It is something to remember when planning for your retirement as the credit cards companies charge 12 to 18% monthly fees (unless you have negotiated a special reduced rate with the credit card company). Over time, the interest fees can become substantial and it is a good financial decision to pay off your credit cards in full before retiring. In general, try to maintain a low balance or 0 balance on your credit cards.

Social Security Benefits

Social security benefits can be collected after 62 years of age. There is no point to collect the benefits if you are still working, but as it is due to you, you should apply for it at 65 years old. If you can spare the income it is a good idea to invest this extra cash. You can deposit it to your RRSPs and get the tax credit for several more years.

Pension Plan

If you have a university appointment or track tenure, you can start to collect your pension at 65 years of age. At certain universities, you can collect your pension a few years before 65 years of age. Your pension will however, be prorated and lower. Interestingly, if you collect your pension earlier, as it is prorated, you will get the same amount over 20 or 25 years as if you started to get it at 60 years of age. This is why starting to receive your pension earlier does not necessarily bring more money over time. But, most importantly, you should get your pension at 65 because you will probably have a better return on your investment.

Estate Planning

Estate planning is a long-range plan aimed at ensuring that your assets are protected, your investment return is maximized and that your spouse and/or your children inherit the maximum amount after you die. The fundamental objectives are to plan adequate funds to cover estate costs, to conserve your assets by decreasing the potential estate income tax and to guarantee distribution of your assets after death.

While estate planning focuses on what happens after your death, you should prepare before your death to meet the goals mentioned above. To that effect, a very important and vital step is to prepare a will. Many people do not realize that if you do not have a written will upon your death, provincial laws dictate many aspects of estate distribution and may create difficulties for your family. It is not only important to draft a will but also to review it every few years, as your assets may change or you may change your mind on the distribution of it due to circumstances.

I recommend the input of a lawyer who specializes in estate planning, so after assessment of your wealth, he can work with your accountant to establish your tax liability and your financial adviser to maximize your return through appropriate investment vehicles.

RRSP

Many people do not realize that a dollar in a RRSP does not have a full dollar value. You have contributed money every year to your RRSP savings and have benefitted from the tax deduction on your personal annual income tax report. Over the years, you have accumulated a sizeable asset. That was the government intended goal but as one starts to withdraw investment return or capital money from the RRSP fund, the government will be taxing this revenue according to your withdrawal and income level. It is a deferred tax program.

There is a good chance that the average physician could live 20-30 years after retirement. Take this into consideration when reassessing your plans. Do not plan to spend all your wealth by a certain age and then find out that you are still healthy. It would be terrible to run out of money at the most vulnerable time in your life when you may need extra help and support.

Create New Interest

Retirement needs to be seen as a new chapter of your life. We have spent most, if not all, of our life identified as a doctor and we may have merged/amalgamated our personality with our profession. When we retire, a person may have a challenge to adjust to this new phase in their lives and need to recreate a new identity for themselves. It is a valuable step, during your career to develop hobbies and interests that you can take with you during your retirement. The next chapter will elaborate on this topic.

PART 5: Family and Retirement

CHAPTER 24: Retirement

Retirement can be a wonderful time in your life. This special phase in your life may include hobbies, travel and volunteering. During this time you may also be employed on a contract basis or working part time. But one of the difficulties with retirement is that many aspects of your life will need to be rearranged. In addition to the obvious occupational, familial and social adjustments, your income will most likely change drastically. Pension plans are not common for doctors except if you are working in a university setting or earning a salary with a major organization.

Changes, Changes, Changes

With these changes to your income, new tax rules may come into play and your investment portfolio may need restructuring. Life insurance can be used for a distinct purpose and your estate may require attention and reorganization. All of these differing aspects to this new phase of your life including; financial, work and hobbies, as well as personal elements are interconnected and a minor change in one part of your life may have a larger implication in another part of your life. Even though planning is an important part of retirement, most people do not make plans before retirement and are not prepared for it as they could be.

Most Canadian doctors are able to start retirement in their early 60's. The new approach to retirement can come in different types of retirement. More and more, the goal of retirement is really about achieving financial independence. This is the point in life where your career and lifestyle choices are no longer driven by financial necessity (for some this may occur before retirement).

Your retirement income depends on how much you save and how you manage your money. Sometimes there can be unfortunate events like the year of 2008 when many people lost money and a good portion of their investments.

Retirement is really about the 3Rs: Relationships (family), resources (money) and re-inventing yourself (personal interests). A perfect retirement would be an equally satisfying and happy balance between all of the 3Rs, but you need to define each of the R's carefully. Define what each of the R's means to you.

Relationship and Family

The longer a couple is married, the less likely they will divorce, even if they suddenly have serious marital problems. After many years of marriage there can be many incentives to stay together (rather than divorce). Some long-term couples will stay together for the extended family. Couples who don't get along usually avoid each other, rather than divorce.

But it's a lot easier to avoid an annoying spouse when you or your spouse is busy working. As many as 60 hours a week can be lost to employment. The rest of the time can be spent doing separate activities. Eventually, the spouse is doing anything other than spending time together.

Retirement, of course, will bring all of that to an end. At this time, some people may find themselves face-to-face with someone who they haven't gotten along with for years. Some blame it on retirement, but the truth is, you are now forced to do something that should have been done from the day you were married: create a lifestyle that takes each other's feelings into account.

But even marriages that have been relatively good often struggle somewhat during the transition to retirement. Now they will be spending most of their time together. That's because retirement changes so many aspects of life. Many couples feel that the solution to their initial discomfort together is to spend less time with each other.

You need to communicate your hopes for the future with your spouse. For better and for worse, retirement imposes major changes on a marriage and change is always stressful. Ending a career, especially one that has been rewarding, is a major life transition. Add to this, a tremendous lifestyle change. Some may consider these changes a loss, but I believe you should consider it a positive new period in your life.

While many people are diligent about saving and investing their money, few consider the psychological effect that usually comes at the end of a career.

Interestingly, this life transition doesn't seem to affect women the same way it does men, even those who've enjoyed a long and fruitful career.

The end of women's working life is often less traumatic because women tend to have multiple interests and work is just one of the many fulfilling things in their lives. At the same time, the majority of women can experience anxiety about their husband's post-retirement life.

Even though a large percentage of the men said they were looking forward to retirement, the free time and enjoying their hobbies, almost none have considered how all of this extra time together with their spouse might affect their marriages.

For some unhappy-together couples, the problems start when they don't have the same expectations of retirement. This can get exacerbated when they don't talk about it. If you want a quick measurement of how compatible you and your spouse are, try being together 24 hours a day, seven days a week. That's the test that couples usually take immediately after retirement. The years that a husband and wife have spent creating independent lifestyles can come back to haunt them on that day, because they may have to face the fact that they have little in common.

Throughout their married lives, they failed to create common interests -- they did nothing to create compatibility. Rather than building a relationship based on mutual respect and sensitivity, they had ignored each other's feelings, missing out on a lifetime of marital happiness. Couples that have grown to become incompatible react to spending time together the same way many retired couples react to retirement. They hate it at first. But as they learn to avoid habits that hurt each other and understand how to meet each other's needs, this time together becomes easier and more fulfilling.

For some people, this is a long-awaited time for new adventures, to revive or create deeper connections with loved ones and discover a new purpose in life. For others, it means a lot of time relaxing: in the hammock, at the computer or on the golf course. To not drive each other crazy, couples need a mutually acceptable game plan for the future. They need to think about and discuss how they want to spend their time, including how much time they want to spend together. These talks should begin long before retirement. It's important to acknowledge the gender differences. Most physicians have made their careers the primary focus of their lives. It was how they measured themselves against others and was the main source of their self-image. Some admitted that they felt less valued if they were no longer bringing in money; others were clearly apprehensive about doing "nothing" for a while. I personally thought I did not know what clothes to wear as I spent my entire career wearing a suit, and I had to put them away and replace them by something else. It took me one year to figure how to dress in a more casual way with sport clothes, casual clothes and sometimes wearing a suit.

Reconnect, Rediscover

Two decades ago, a Japanese physician found that as many as 60 percent of wives of Japanese retirees were suffering from similar physical symptoms, which included; depression, tension headaches, stomach ulcers, rashes and other signs of stress. He dubbed this "retired husband syndrome," and researchers speculated that the women were becoming ill because their retired husbands were treating them as if they were still the boss. You may not treat your wife/partner as if you are the "boss", but being home a considerable amount of time can certainly put a stress on the relationship. This is why communication is critical.

Not all men and women find the transition to their own retirement easy. Some report feeling like they were playing hooky if they visited a museum in the middle of a weekday. Others find it hard as many or several of their friends are still working or are absorbed with grandchildren.

Whether you and your partner's plans for the future are 100 percent on the same page or totally out of sync, these suggestions will help you create a balancing act and happier future.

1. Take time to adjust to being retired. You don't have to do everything you've been planning the first month. Be patient with each other, especially if you want to get up and go and she/he wants to sleep in without an alarm clock for a while.
2. Express yourself if your partner wants to do something that you don't want to. If you are used to accommodating your spouse just to keep the peace, it is never too late to change that behaviour.
3. Stay connected to others by taking classes, joining clubs, volunteering or being involved in your local community.
4. Negotiate sharing more household responsibilities. If your partner didn't help with these duties before, suggest she/he start with the things you dislike the most and take it from there.
5. Stay active. Exercise, play sports, go to the gym and consider beginning each day with a walk and a coffee with your spouse. This can be good for your health and can be a nice ritual for discussing your plans for the day.
6. Make sure you each have enough "alone time." It is perfectly reasonable to have your own activities or see friends on your own. There is nothing wrong with a girls or a guy's night out. It can be refreshing for a relationship to take time away from each other once in a while.

Try not to over plan or over schedule. Leave time to do the unexpected or to just hang out. Enjoy your time rather than being stressed by it.

Resources and Finance

1. Re-invent your job:
   Before you even think about giving up your full-time job, you need to figure out where your retirement income will come from and how much you will need to support your life style. If you don't have enough money to retire, then you'll need to make some tough choices. You can cut back your planned retirement spending, or find a way to save more. But these days, many Canadians are choosing to work longer. The average retirement age is 62, but that is changing and we can expect many doctors retiring in their late 60's.

   Working longer doesn't have to mean working full time. As a doctor, you have an array of part-time, contract and temporary jobs available. You may plan a few years earlier with your group practice to decrease your workload and time, or find a part-time locum for your family practice to allow free time. You will have to accept less than what you previously earned at the peak of your career, but even a modest retirement wage can have a substantial financial impact.

   If you don't like the idea of working longer, you will need to increase your savings and reduce your expenses. This takes discipline, but even if you've saved a little by your late 40's or early 50's, you have enormous potential to save if your mortgage is paid in full and your children are financially self-sufficient. The idea is to redirect the money that used to go to your mortgage and kids into savings; you can probably save more than 30% of your income (counting RRSP refunds) if you set your mind to it. Do this steadily for a few years and it can add up to a considerable sum that may allow you to work only if you want to.

2. Investment income:
   More than 75% of Canadian physicians earn a fee-for-service income. Retirement income for your investments will need to be planned carefully. While fixed-income investments can protect your savings, you're not likely to grow your wealth with GICs and bonds. To stay ahead of inflation, you will need to keep a significant portion of your portfolio in equities and focus on dividend-paying stocks which may provide the right balance of risk and reward. The best way for people to get a decent return these days is to have a good portion of that return come from reliable dividends. Picking the right dividend stocks is key. Please consult the chapter 14 "Investing" for more information on this.

3. Cash in on your home:
   Many people approaching retirement are afraid they may not have enough savings to retire with, but there is one area, which you can capitalize on if you own a house. Real estate in Canada has enjoyed an enormous boom in recent years and this has allowed many homeowners to build significant wealth. This newfound wealth can give you more options in retirement.

   If you own an expensive home, you can add to your cash savings by downsizing or relocating. For example, some Vancouver homeowners are selling modest-sized homes, as the real estate market has been highly valued for many years. They follow this up by buying a two-bedroom condo in town or a smaller home nearby, and winding up with $500,000 in their pocket. It's more common for homeowners in other parts of Canada to net $100,000 or $200,000 after costs, as some areas of the real estate market has not plus-valued as much.

   "Phil and Brenda, retired couple in their 60's wanted to sell their modest house and move to a small town and be closer to their children. So they sold their house for $980,000 this spring and bought a renovated detached house in a desirable small town near the ocean for $620,000. They netted around $300,000 after costs, while retaining home equity. They invested their money into their portfolio, increasing their investment return."

In your retirement, if you run low on savings the equity in your home can be a back-up plan. If you stay put, you can cover essential expenses for as long as possible with your investment income and then if you need extra cash, you can borrow against your home equity with a reverse mortgage or home equity line of credit—albeit this should be a last resort. Later in life, if you move into a retirement or nursing home, the proceeds from selling your house can carry those costs for several years. Even if you never draw on your home equity, it can provide a great legacy for your children.

4. Think differently about debt:
   Carrying debt into retirement was once considered dangerous and irresponsible. But today's low interest rates have changed the game—as long as you borrow in a smart way: getting the lowest interest possible, ensuring your revenue or investment bring a return sufficient to pay your debt and keeping your debt at a manageable level. You can do that by making sure your debt payments are much lower than your capacity to cover them.

   You can take advantage of today's low interest rates and this could potentially help to put you in a better position later in life. Many Canadians may soon renew their mortgages; this is an opportunity to obtain a low remortgage interest rate and lock it in for 5 to 10 years. When your current mortgage interest rates are one or two percentage points higher than what you might be able to negotiate today, it can be substantial. On a $200,000 mortgage, that's about an annual savings of $2,000 - $4,000, you can use this extra money to make extra mortgage payments or, if necessary, pay off other debts. If you have high-interest credit card debt that you can't seem to pay off, you might consider using your home equity for a consolidation loan. The consolidation loan can provide you with much lower interest rates and help you pay off your credit card debt (with higher interest rates).

Just remember that your main goal is getting rid of that consumer debt, not just making it more manageable. Debt can be seductive, but as you approach retirement it's critical to only borrow for productive purposes and for as short a period as possible.

5. Wait before you buy an annuity:
   Many retirees like the idea of annuities, which provide guaranteed income for life in exchange for a lump-sum payment. In the past, people often purchased annuities as soon as they retired in order to replace their employment income, especially if they had no pension. That may not be such a good idea any longer. Annuity payout rates can be affected by interest rates and current payouts are dismally low.

6. Reduce your tax bill:
   Before 2009, RRSPs were really the only way for Canadians to shelter their retirement savings from taxes. But the introduction of the Tax-Free Savings Account (TFSA) has added another option.

   Unfortunately, the rules are complex and it's not easy to figure out how to combine these two tax-sheltered investment accounts for your maximum advantage. If you have high income today, it makes the most sense to use RRSPs first, since you will get a larger tax refund, but since the total amount allowed per year for TFSA is low, your income should allow you to save the maximum in both tax shelter methods.

   The reasons are several but most importantly, the TFSA is more flexible. If you need the money for an emergency you can withdraw TFSA money without tax consequences, whereas RRSP withdrawals might cause you to pay hefty taxes if you are still working. TFSAs are also better if you expect to end up with sufficiently low income in retirement and be eligible for the Guaranteed Income Supplement (GIS). Withdrawals from a RRSP reduce GIS payouts, whereas TFSA drawdowns do not.

You will need to figure out how to withdraw money without paying too much tax. If you have a substantial amount of RRSPs and non-registered accounts, it can become even more complicated. The best strategy is to take a balanced approach to withdrawing money from all your sources. You will want to keep in mind our progressive tax system (where higher incomes get taxed at much higher levels).

Seniors who defer RRSP withdrawals until 70 years of age are often forced to make large withdrawals after age 71. This is when they are required to convert RRSP funds into a RRIF or an annuity. This can often push them into a higher tax bracket.

Therefore, you should calculate the income you will require in relation to your lifestyle needs. Then maximize your income tax bracket by splitting income between dividends from your corporation, income to your spouse, income from yourself and your spouse's RRSPs.

The path to financial independence isn't always easy. If you set reasonable goals and make adjustments to stay on track, it will give you more freedom later on. Then chances are you'll achieve the retirement plan that suits you best.

Reimagine (Interest and Hobbies)

If you don't do anything but work, eat and do a little exercise and live a sedentary life, you will need to develop a few outside interests. Hopefully you will do this before you retire. Retirement hobbies should be fun and relaxing. Hobbies don't need to be physical unless you like physical effort. Golf and other competitive sports are great hobbies as long as you don't take yourself too seriously. Do not consider your hobby as another job but rather an activity that you can enjoy alone or with friends. A great hobby should assist in managing stress and not create more stress. There are ways to find an enjoyable hobby that brings both relaxation and social connection to your life.

Retirement should not be considered a conclusion to a person's life but rather a new chapter in life. This phase of your life will certainly be different than the previous chapter in your life. And there will be all kind of new opportunities available for you to try.

Experience brings knowledge and hopefully, wisdom. Without the burden of a daily job, you have time to consider the memories of past people, work, events, and places. Retirement allows you to recognize your accomplishments, understand and forgive your perceived failures and set a new course for achievements in your life.

To provide some guidance on retirement, I will list a few activities. These are only a few examples. You may learn about additional activities that you find entertaining and enjoyable in this new phase of your life.

1. Working:

   Some people may have difficulty to stop working "cold turkey." You may want to consider scaling back your work to part time, taking on some consulting or seasonal work, or otherwise find a work schedule that also offers plenty of time for leisure pursuits. Many doctors have the possibility through their practice to discuss pre-retirement with their colleagues and plan to go off-call for instance, to go down to 50% workload over a certain period of time before full retirement.

   "Jim has planned for his next 5 years with his group. He will give 25% and money of his time for 2 years, 50% for another 2 years and then full retirement. His group has agreed and will supplement the extra workload with locum."

2. Sharing:
   You have a lifetime of accumulated memories and experiences. Consider writing a book about your experiences, about an aspect of medicine that interested you (which is what I am doing). Put your thoughts on paper, wrestle with their meaning and anticipate the opportunities it may offer. For a more interactive experience, consider starting a blog about a topic that interests you. It is easy, very interesting and can be challenging and stimulating.

3. Learning:
   You can decide to learn something new. Retirement is the perfect time to acquire new skills, pursue a new hobby, or just learn for your own interest. Some colleges and universities even offer free or deeply discounted tuition to retirees above a certain age.

   "Mia retired at 52 years of age, at the same time as her husband Nick. She decided to enroll in a Contemporary Psychology Group Study, for her own interest and not only draws much pleasure from it, but also created a new group of friends."

   Reading is a very enjoyable and relatively inexpensive hobby in relation to other forms of entertainment. Many of us will finally have time to pick up a book. To be able to read and simply learn about something new. It does not have to be a medical book or a medical journal.

4. Spending time in another country:
   Being able to retire after a lifetime of work is quite an achievement. You may plan to celebrate the occasion by purchasing a small condo or house in a warm winter area, such as the Southern United States, Mexico, Central America or Southern Europe.

   "Mitchell and his wife sold their house, bought a smaller condo in their town and bought a small house in Palm

> Springs. They remained with a small affordable mortgage but enjoy spending the rainy and cold winter away in sunny California."
>
> Just make sure you can truly afford the extravagant purchase and won't risk running out of money too soon.

Applicable to owning a house in the USA, which is a common place for Canadians to live in the winters (hoping to avoid the cold weather), you should be aware of three important points. If you have a property somewhere else in the world, these three points may not apply, but you should look into what rules might be in place at the new location you are residing in.

First, the Entry/Exit Initiative of the Perimeter Security and Economic Competitiveness action plan will exercise a tighter control on how long Canadian and American travelers stay in each other's countries. Passports will be swiped at entry and exit points on both sides of the border. Both countries share data, which means that the US and Canadian authorities know exactly how long you stay in the respective country. If you were not exact with the calculation of your days across the border and thought it did not matter, now it does and you can be caught, which can result in trouble with immigration and the IRS (the American Tax Agency).

Second, a legislation has been proposed but not enacted as yet, as part of the US JOLT (Jobs Originated through Launching Travel) Act, which will allowed Canadian retirees to stay up to 8 months a year, rather than the current 183 days, as long as you are over 55 years of age, continue to maintain a home in Canada and will not be working or applying for social assistance in the USA. But at this point, it is not in place and even if it was, JOLT falls under immigration and there has been no indication that the IRS will agree to this change or make appropriate adjustments to align the 8 months allowed under JOLT, with the 8 months allowed by the IRS before filling an income tax report.

Third, the limit we understand of 183 days stayed in the USA, can actually be less because of the States substantial presence rules.

The rules are that if you are in the US more than 30 days in any given year, you must add up that number of days, one third of the days from the preceding year, and one sixth of the days from the year before that. If the total amounts to more than 183 days, you are deemed US tax resident.

For example, if you go to your vacation home in the USA for just under 183 days per year every year, the calculation of adding 1/3 of the days and 1/6 of the days for each previous year, will put you over the 183 days allocated per year. If you are spending more than the allowed 183 days per year in the US, you could be deemed a US tax resident. You can however, avoid the substantial presence rules by filling through your accountant the Closer Connection Form (currently form 8840) with the US IRS which evokes the US-Canada NAFTA Treaty and declares you to have a closer connection with Canada.

Please consult http://www.grasmick.com/snowbird.htm Enjoy a nice long warm vacation.

Travel:

While you are working, a desire to see the world must be fit into carefully planned vacation days. Once you are no longer tied to your job, you can plan to travel as much as you want in relation to your family, budget and world interests. You have more freedom to travel during off-peak times (and more affordable times) of the year and to stay as long as you want.

You can learn a new skill in your travels such as; gourmet cooking (in Tuscany), participating in either domestic or international humanitarian projects, you can tailor your excursions to fulfill passions like antiquing, museum hopping or just lazing around on the beach. Whether it's a weekend getaway by car (or bicycle) or a cruise around the world, a change of scenery of your choice is an amazing feeling when the week doesn't start with a long commute and a quick cup of coffee before the Monday morning meeting.

"Through his hospital relationship, Mike and his group had done work at a children's hospital in Shanghai, China for 8 years. When he retired and the Chinese project was completed, Mike used his contact at the hospital to develop a new relationship with a children's hospital in Chennai, India and carried out volunteer work for several years."

"Jim loves cooking. Finally, in his retirement, he was able to organize cooking lessons in the vineyards region and attended a full week of cooking lessons in Tuscany. He continues to attend cooking lessons in town."

You can even be more adventurous and experience another culture. Once you're no longer tied to a job, you can truly live anywhere in the world. You do not need to leave your family and friends but you can plan to spend some of your retirement years living abroad.

"Rod and his wife had always wanted to spend time in Provence. They achieved 3 goals at once. He did an exchange of his house with a professor in a small town in Provence, he spent one year learning French cooking and his wife learned painting. It was so rewarding that he is planning on another exchange, this time in Italy."

For keeping travel cost down, try to remember the possibility of a house exchange. Through your network of university, friends and colleagues you can find a colleague/ professor/friend in another country who is interested to exchange the use of their home for the use of yours.

Family:

Many retiring doctors are embracing home and hearth. Some are just starting to become grandparents and that's motivating many to incorporate more family dinners and parties into their regular activities. While you were working, you may have missed many of these family gatherings (being on-call) and now is the time to spend quality time with your loved ones. Travelling as a family on a cruise, or renting a summer cottage, for example, are becoming popular family activities.

Hobbies:

Activities that doctors could not fully enjoy when working full time, become more interesting when retired. From cooking to painting, to playing an instrument, many retired doctors pursue leisure interests that offer physical and psychological pleasure. The outdoors can be a big draw and gardening can an extremely therapeutic activity. Whether planting a row of corn or flowering plants, gardening is a hobby enjoyed by many retirees. It also allows for creativity and burns calories. An interesting fact, more people fish than play golf or tennis combined. Many travel agencies are creating fishing vacation packages that allow people greater flexibility to enjoy this recreational sport year-round. Older golfers will be in full swing during retirement. To be closer to the action, many retirees are buying homes in communities built around golf courses.

"Patrick has been playing violin for pleasure for quite a long time. He and his wife love to go to classical concerts and the philharmonic orchestra in their hometown. In retirement, Patrick found new pleasure in directing the college cello musical group."

Whatever hobby you are interested in is the important point. As a doctor, there aren't any specific hobbies that you 'should do', or you feel you should do. We need to break down societal stigmas around retirement and participate in, create and learn what we enjoy. You will be rewarded with a feeling of joy, peace and satisfaction.

Volunteering:

Donating your time or skills can be satisfying - especially among doctors whose social awareness encourages them to be connected to their community and organizations. Their involvement takes many forms, from sitting on an organization's board to teaching overseas.

You may be interested in donating some of your time by doing volunteer work for a charity or community organization. Make sure to find a volunteer position that matches your interests and the skills you have acquired during your working life. Volunteer efforts don't have to be on a grand scale either. Whether retiring doctors are contributing by helping animal shelters, organizing clothes or stocking local food banks to aid the hungry, they are offering a helping hand where it is needed.

Work is how we create our identity as individuals and how others know us. To continue this sentiment of individuality, you can volunteer your time, your years of experience and wisdom. This can help you fulfill your need to feel productive and be recognized for your efforts during retirement.

"Mario has always been a computer buff. He always buys the latest Apple technology for his office and himself. He has also attended a few Apple trade shows in San Francisco. When he retired, he had a friend who managed an Apple store. His friend brought Mario in to help at the store. Help turned into an opportunity work as a part-time sales person, working 2 days a week at the store. Mario stays abreast of the latest technology and loves to talk with people, especially older people because of his patience. He invariably ends up selling them new Apple technology."

"David has always been a great doctor. He and his wife are absolutely lovely, always trying to help others. David had found himself (through friends), working for a funeral home and driving the hearse. He likes to talk to people even during their difficult emotional moments, enjoys driving the hearse and making sure the bereaved family has as good an experience as possible."

"Ron has travelled quite a lot to give lectures, do manuscripts and research work presentations. He had accompanied several missions organized by his hospital and department. When he retired, he took a special interest in helping a young surgical colleague and volunteered his time to visit his colleague in India and a nearby orphanage with his wife."

Exercising:

Exercise is one of the biggest activities many retirees will embrace as they become more aware of its importance. Retirees can no longer claim they are too busy to exercise. In growing numbers, retirees are taking their training to a new level and joining health clubs. Playing tennis or going for a walk not only serves a physical purpose, but a social one, too.

So when you exercise, or if you are starting an exercise program, use common sense – don't exercise when you're ill, wear bright clothing when exercising outside and learn the difference between pain and discomfort. If you lift weights, pay close attention to the execution of the exercise, avoid difficult or painful motion, and don't hesitate to ask for help. The same applies for jogging, bicycling, tennis, and golf. You should exercise in moderation except if you are a fit marathoner or cycling a long distance. You will still enjoy it and will protect yourself from injury.

Retirees who take part in regular physical activity are privy to countless health benefits, including lower body weight, greater strength and endurance, increased flexibility and balance, and better mental health. In fact, you'd be hard-pressed to find any research suggesting the older you get, the less active you should be. Being in shape doesn't mean achieving washboard abs or running marathons. It means being able to take your children and grand kids on a ski vacation or being able to take care of daily household chores without experiencing exhaustion.

Retired women are staying active and keeping fit, and although sports like golf and bowling are definitely on the list of popular leisurely physical pursuits, yoga offers a workout, as well as mood enhancing and maybe even bring spiritual benefits.

Home Improvement:

With the kids out of the house, and with better financial resources, it might be time to give the kitchen or other areas of your home the makeover you have always dreamed about. You can add quality touches to your home such as granite countertops or wood cabinetry. Tackling a project can be extremely gratifying, especially if you have always worked behind a desk. If you are handy, you can save money by tackling home improvement projects yourself. You can also get some exercise and beautify your home by taking up gardening. Whatever you do, enjoy the process and do not take on too large a project that could potentially create stress or cost more than expected. If you have difficulties you can always ask for assistance from a reliable contractor. Keep it manageable, simple, low stress and pleasurable.

Conclusion

The average person has roughly 20 to 25 years of life after retirement – plenty of time to write a book, run a marathon, or mentor young colleagues or students. During retirement, there is even time to do nothing. You can discover the beauty of grandkids or enjoy the history of a long relationship. Tomorrow can be the beginning of new adventures, new joys, and bigger successes – how you spend your retirement time is up to you, as you are your own master.

Acknowledgments

When I retired, I was not sure what the future would hold. Travelling, exercising, bicycling, teaching overseas and what else I did not know. I was not prepared for how the pace of my life would slow down. Despite being active, with this slower pace, I still felt that there were too many hours in the day. I quickly came to realize that I was not as prepared for retirement as I had believed.

Along with the unwavering support of my wife and the insightful questioning of Dr. Rory Gorshon, psychologist and friend, the idea of writing a book on my professional experience became a reality.

When I was growing up, I decided to go to college, then university and moved on to a surgical residency. At that time, there was very little help for people entering the medical profession. Finding a position after my residency was not easy and establishing my practice was a challenge as well. Surrounding myself with amazing support was more about being at the right place at the right time, than because I had the knowledge on how to do this. Everything fell into place because of good timing.

Reflecting back, I realize that what made my career successful was not only my surgical skills or my decisive personality, but equally the people I surrounded myself with. My secretaries, accountant, financial advisor, banker, lawyer, travel agent and my closest and respected friends were all a part of the support system, which allowed me to be able to function successfully. Without their support over the years, this book would not have been possible.

This book is dedicated to those people who supported me over the years, my closest friends and colleagues. Dr. Phil Ashmore, a pediatric cardiac surgeon who was not afraid to believe in a young surgeon just out of training. Dr. Ashmore was the most amazing mentor, a very special friend, and a father figure. I will never forget. Dr. Robert Freedom, esteemed pediatric cardiologist, who provided sound advice when I started my practice. Ms. Charlotte Robertson, my secretary who spent 27 years doing an amazing job at keeping me out of trouble and was integral to the smooth functioning of my practice. Ms. Joan Wilkinson, who joined the office later and who was a valuable asset to my team.

Dr. Peter Edmunds, who was my right arm and single surgical support for two years and remains a most esteemed friend. Dr. Guy Fradet, cardiovascular and thoracic surgeon, was always there when I needed surgical advise and through challenging times. His long-term friendship going back to medical school has endured. Dr. Fred Kozak, ENT surgeon, who never failed to listen, to question me, and remains a great friend.

Mr. David Melloy, my trustworthy financial advisor who lead me to financial freedom. Mr. Gary Huebner, financial adviser, who gracefully accepted to review the investment chapter of this book. Mr. David Rolfe, an informed solid and strict accountant who kept me out of trouble. Mr. David Cender, my US tax accountant and friend, who provided insightful advice on complex tax issues and reviewed the accounting chapter of this book. Mr. Bob Cowan, insurance adviser, who reviewed the chapter on insurance, a great friend having provided valuable career advice. Ms. Jenny Sameshima, travel agent, who allowed me to travel the world with ease and always ensured that my travels were very well organized. To Ms. Catherine Douglas, Rogers Media and Maclean's Magazine, for her assistance and support for my book. To my dear friend, Raj (Chennai, India), Ms. Charlotte Kapitza, Mr. Keith Bockhold and Ms. Sacha Devoretz, for their invaluable contribution in editing my book.

To the wonderful people of the Cities of Toronto, Canada, and New Orleans, USA, my homes for a wonderful period of my life during my residency in Adult and Pediatric Cardiovascular and Thoracic Surgery. The people in these cities helped me grow as a physician in training, as a person and as a future clinician. To the many people of British Columbia Children's Hospital that I was lucky to work with, to learn from and provide such a valuable support in helping me care for the children of British Columbia.

To Susan, my dear wife and partner, a person I was so lucky to meet, marry and retire with. Life has never been so good and I am still learning.

To all these people, I am forever indebted for such a successful career. Thank you.

About the Author

Dr. Jacques LeBlanc has been an adult congenital and pediatric cardiovascular and thoracic surgeon at British Columbia Children's Hospital, Vancouver, British Columbia. He retired 4 years ago after 30 years of practice.

He was instrumental in developing multiple areas of pediatric cardiac surgery and adult congenital surgery in Vancouver, quality assurance and a surgical database for the pediatric cardiovascular and thoracic program while remaining at the forefront of technological advances. More recently, his work experience involved teaching overseas, directing a team of colleagues and health care professionals in the training of staff at the Fudan University Children's Hospital of Shanghai. He was personally involved in teaching and training the pediatric cardiac surgical staff at Apollo Children's Hospital of Chennai, India, over the last 4 years.

Dr. LeBlanc has co-authored a medical book, "The Operative and Postoperative Management of Congenital Heart Defects" published in 1992. He has published 74 medical articles and made 96 verbal presentations.

He currently lives in West Vancouver, BC.